Pain-Free Biochemistry

Pain-Free Biochemistry

An essential guide for the health sciences

Paul C. Engel
University College Dublin, Dublin, Republic of Ireland

WILEY-BLACKWELL

A John Wiley & Sons, Ltd., Publication

This edition first published 2009
© 2009 by John Wiley & Sons Ltd

Wiley-Blackwell is an imprint of John Wiley & Sons, formed by the merger of Wiley's global Scientific, Technical and Medical business with Blackwell Publishing.

Registered Office
John Wiley & Sons Ltd, The Atrium, Southern Gate, Chichester, West Sussex, PO19 8SQ, UK

Other Editorial Offices
9600 Garsington Road, Oxford, OX4 2DQ, UK
111 River Street, Hoboken, NJ 07030-5774, USA

For details of our global editorial offices, for customer services and for information about how to apply for permission to reuse the copyright material in this book please see our website at www.wiley.com/wiley-blackwell

ISBN: 978-0-470-06045-2 (HB)   978-0-470-06046-9 (PB)

A catalogue record for this book is available from the British Library.

Typeset in 10.5/12.5pt Times by Aptara Inc., New Delhi, India
Printed in Singapore by Markono Print Media Pte Ltd

First impression   2009

Contents

Preface

This book is written in the hope of filling what I have felt as a teacher to be a glaring gap in the availability of books on biochemistry suitable for students of nursing and related healthcare subjects. Textbooks for medical students are relevant but far too detailed and offputting for students on a short course. Standard textbooks of biochemistry for science students, although there are many good ones, are not only too detailed but also misleadingly broad. The patient is unlikely to be an insect or a plant, and so the biochemistry of locust flight muscle or spinach chloroplasts is only of limited help to a hard-pressed nursing student. On the other hand, the various books on science for nursing are usually very sketchy when it comes to biochemistry.

At first encounter, biochemistry can be quite intimidating. I have therefore tried to strip away as much scary detail as possible, leaving out what I felt to be unnecessary structures, enzyme names and reaction steps, and keeping in only the bare minimum to allow some genuine understanding of what is going on in metabolism. I am sure students will find this quite scary enough! There are a large number of topics I have not touched upon, e.g. the intricacies of signalling, inflammation and molecular immunology. What I hope is that I will give a student a firm enough grasp of what biochemistry is all about to read up about anything else that they specifically need in their own speciality.

As my own experience comes from teaching nursing students in Ireland, where many arrive without a chemistry foundation, I have also interspersed some basic chemistry. This is in honour of the intrepid group of mature women students who pinned me against the blackboard some years back and said 'Professor Engel, we can tell that you're enthusiastic about these protein molecules you keep mentioning, but what you don't seem to realise is that we don't know what a molecule is!'

I need also to thank Margaret Deith who read it all with the combined eye of student, teacher and editor; my wife, Sue, who kept reminding me that not all students have the same learning style and pointing out where what I thought was obvious might not be obvious to a student; Carmel Nolan who read it from the point of view of a lecturer in the School of Nursing, spotted obvious gaps and encouraged me to keep going; and Francesca Paradisi who checked the chemistry sections for any glaring errors. Finally, Nicky McGirr, my commissioning editor was ever helpful and enthusiastic, Fiona Woods, Izzy Canning and Robert Hambrook at Wiley in Chichester guided me patiently through the final stages and Lesley Montford handled the copy editing with efficiency and understanding!

Visit the book's companion website at www.wileyeurope.com/college/engel Here you will find explanations of the answers to the MCQs contained in the book plus an extended set of MCQs ranging more widely across the topics.

Paul Engel
July 2009

SECTION 1

Foundations

TOPIC 1

Why biochemistry?

Aims of biochemistry

'Why do we need to learn this stuff?' is a question that every biochemistry lecturer is likely to have met over and over again in teaching healthcare students. It is somehow easier and more obvious to see why you need to know what the heart and kidneys do, where the veins and nerves are, etc. than how and why we metabolise glucose. So do you as healthcare professionals really need to bother about biochemistry?

It is now possible to provide a reasonably good broad description of what is going on at a functional level in most of the processes of the body, e.g. fertilisation, growth and differentiation, inheritance, immunity, **hormone** action, nerve conduction, muscle contraction and so on without biochemistry, but all these accounts end up prompting a degree of disbelief and the vigorous question 'How on earth can that possibly work?' Each of these stories invokes shadowy objects and influences that seem to have amazing powers – substances that recognise an invading **bacterium** or **virus** and target it for destruction, substances that switch on a program to turn a girl into a woman or substances that allow something happening inside my head to move my fingers across my keyboard! Biochemistry recognises that none of the marvels of biology can be properly understood without getting right down to the chemical level. Over the past hundred years or so this discipline has emerged by developing and refining chemical and physical tools to study very complex molecules and systems and has achieved an in-depth understanding of many of life's processes that, even 30–40 years ago, were entirely mysterious.

Pain-Free Biochemistry   Paul C. Engel
© 2009 John Wiley & Sons, Ltd

Health care and biochemistry

Even so, do *you* need to understand at that level? Leaving aside the thought that you might be genuinely interested and curious to find out, there are many other pressures on your crowded learning time. We should therefore recognise that the people you will look after will usually be under your care because something has gone wrong! Even healthy people coming in for a check-up need clinical staff who can recognise the early signs of a problem. These days the check-up is likely to involve a battery of 'tests', which typically involve sending blood and urine to 'the lab'. Some of these tests may involve microscopic examination, e.g. to recognise abnormal distributions of blood cells, but very many of the tests are biochemical, and 'the lab' is the hospital's biochemistry department (possibly under the label of chemical pathology). If indeed there is a problem, patients ideally want you and your medical colleagues to offer something more than sympathy. Accurate diagnosis and effective treatment are what they are after – in much the same way that, if there was something wrong with their car, they would also hope for accurate diagnosis and treatment!

All the same, could you not leave all that detailed biochemical knowledge to the physicians and surgeons? Here we need to think about where the analogy with getting a car fixed falls down: when a car is fixed the owner collects it at the end of the day and drives it home; when he or she takes his or her body to be fixed, the chances are that he or she will have an aftercare period in the hands of nurses, physiotherapists, etc. It will be the nurses who deliver the drugs, set up the drips, take the blood samples, change the dressing and supervise the feeding. The patient is going to have far more contact with the nursing staff than with the surgeon and therefore will hope that the nursing staff understand what they are doing!

In considering how far biochemistry is essential to that understanding, let us briefly touch on four disease conditions, two of them longstanding killers and the other two more recent, but all four in their separate ways a source of fear for the public, and see where biochemistry fits in. The first two are the main killers in Western society: cancer and heart disease.

Cancer

In 1972, US presidential candidate Richard Nixon announced that, if elected, he would devote $100 million to curing cancer. The clear implication was that, in its pursuit of frivolous curiosities, the irresponsible, self-absorbed scientific community had overlooked this major source of public concern. It did not occur to Mr Nixon that the real problem was that we could not launch a full frontal attack on cancer until we understood what it was. Of course, at one level we knew that cancer involves our own cells growing out of control, and, knowing that, it was already possible to attack cancers by surgery, chemotherapy and radiotherapy. What we did not know, however, was what led a particular cell or set of cells to become cancerous, and scientists were hotly contesting various rival theories – cancer was caused by viruses, cancer was inherited in our **genes** or cancer was provoked by dangerous chemicals, **carcinogens**. These were all advanced as mutually exclusive alternatives. Today, we understand that all these factors play a part and that a cell has to take several steps, not just one, to become committed to cancerous development.

What had to happen to reach our current understanding was a whole series of discoveries from various unexpected directions, many of them helping us to better describe normality, before we could explain abnormality. The point is that the new understanding is entirely biochemical. No amount of staring down the microscope at stained histological sections of diseased tissues would have brought us to the destination. What was needed was a proper understanding of the various molecular switches involved in cell division, and also an understanding of what controls the development of new blood supply – because once a tumour acquires its own blood supply it becomes more dangerous, seeding new tumours round the body by a process called **metastasis**. Today, with a much fuller understanding of what is going on and of the differences between different tumours, it is possible both to refine diagnosis and to devise better targeted and more effective treatments, so that at least some cancers that were fatal 30 years ago, such as testicular cancer, are successfully treatable today.

Heart disease

What about heart disease? Again it has been known for a long time, at a plumbing level, that the pipes get furred up. A build-up of plaque on the walls of the coronary arteries eventually leads to serious constriction and angina, and, if a blood clot builds up, possibly also to a fatal heart attack. As with cancer, this level of knowledge has provided a basis for surgical intervention of various kinds for a long time, but what about prevention, what about prompt and secure diagnosis, what about long-term medication? And once again, if there are genetic factors, how does this work? Are we all equally at risk? And does it matter what we eat? A proper biochemical understanding is really rather helpful. Today, one of the most widely prescribed classes of drugs is the **statins**, given prophylactically (i.e. for prevention rather than cure) to middle-aged patients, especially men. These are designer drugs specially targeting the production of **cholesterol** by our own body chemistry (see Topic 21), excessive levels of cholesterol now being well understood as a risk factor. Recognising the role of cholesterol and then targeting the drugs to decrease cholesterol levels is based on detailed biochemical research.

AIDS

The other two examples relate to more recent scares. In the 1980s, there was near-panic among various communities over the emergence of a new and deadly sexually transmitted disease, AIDS (acquired immune deficiency syndrome). The first step towards effective action was a recognition that AIDS is caused by the human immunodeficiency virus, HIV. The panic, however, reflected the fact that the condition was so poorly understood that there was no basis for effective therapy. What followed over a very few years was a truly remarkable response by the international biochemical community, which established exactly what HIV was, how it worked and where its weak points might be, allowing us to attack it instead of letting it attack us. This work led on to several strategies for drug design and to the current combined therapies, which allow HIV-positive individuals many years of healthy life rather than a very high chance of debilitating and terminal illness. AIDS, of course, is still devastating many parts of the developing world, but this is now an economic and political problem. The scientific and clinical problem is largely solved.

BSE

Finally, even more recently, in the 1990s, we have had another panic in Europe, this time over bovine spongiform encephalopathy (BSE – 'mad cow disease') and beefburgers. In this case, the successful containment of the problem, which seems to be steadily abating now, has been down to successful public health and agricultural measures rather than clinical intervention. Nevertheless, those measures in themselves have depended on gaining an understanding of an exceptionally baffling disease. It was baffling because it was an infectious disease that challenged all our previous notions of infection – no bacteria, no viruses, just the **prion** protein. In this situation, epidemiology (the study of the pattern of the spread of disease) clearly took us some distance in identifying a link with cattle, sheep, etc. even before we knew exactly what it was in the meat that was so dangerous. This left political dilemmas over the correct way to proceed, as possible methods of prevention struck at people's livelihoods. Deep study at the molecular level since then has helped our understanding not only of BSE but also of various so-called **amyloid** diseases, such as Alzheimer's. The original puzzle is solved when one understands that prions (and similar **proteins**) are molecular troublemakers! They not only misbehave themselves but they also lead their fellow molecules astray. It is inconceivable that BSE could have been properly explained by anything other than biochemical research.

Aims of this book

Biochemistry, then, is very clearly the engine driving major advances in clinical understanding, in diagnosis and in intelligent therapy. It is also a fascinating subject, but it is undeniably complex. Life is complex, and we cannot escape some complexity if we want to understand it. If you want to read *War and Peace* in the original, you have to make an investment in learning the Russian alphabet and language. The effort, one assumes, is finally rewarding. Similarly, biochemistry has its own alphabet and language, which can seem intimidating. My contract with the reader is to try to explain things as simply as possible, to explain the language of biochemistry clearly, and not to introduce complexity for its own sake. I would like to produce a few converts, students with a real sense of intellectual satisfaction in understanding the scientific basis of procedures in everyday clinical medicine and nursing. However, even if I do not succeed to that extent, I hope I will persuade you that biochemistry is not an unnecessary imposition but an essential tool, and that, with a little effort, the main principles can be understood.

This book

Finally, a word then about the organisation of the book. The numbered chapters listed in the Contents are essentially of three kinds and are colour coded as such by title. The majority (blue titles), of course, are mainstream biochemistry, and the individual biochemical topics are presented in short chapters, each making one main point and of a size to be taken in at a sitting. Interspersed, however, especially in the first half of the book, are chapters (red titles) providing some basic chemistry. This is gently presented on the assumption that you may have done very little chemistry, may have found it hard or perhaps did it so long ago that you have forgotten it. On the other hand, if you have a confident foundation of school chemistry, you can probably skip these red chapters. Finally, a third category of chapter, especially in the later stages of the book, presents clinical topics from a biochemical perspective (green titles).

Throughout the book there will also be reference to a series of appendices. Biochemistry is a vast subject, and for most healthcare purposes you probably need only the broad outlines plus certain relevant details. On the other hand, here and there puzzling facts may seem to emerge like rabbits out of a hat. If you are happy to take these facts on trust, you can safely ignore the book's appendices without your patients suffering in years to come. If, on the other hand, curiosity or frustration draws you in further, the appendices are there to help out.

Like all subjects, biochemistry has its own secret language that is very useful once you know it, but a bit offputting till then. Therefore, there is also a glossary at the back for quick reference, to remind you of the meaning of individual terms.

Lastly, students always worry – Have I really understood it? Will I pass the examination? To help check on that there are about 150 MCQs, a few on nearly every topic and immediately following that section, with answers given at the end of the book. The best way to use these is probably not to look at them until a few days after reading the topic. The topics are so short that, if you look at the questions immediately, you are almost bound to get most of them right. If you revisit a few days later, the questions will be more effective in helping you decide what has or has not sunk in.

The book as a whole is intended to help and fill a need. In that spirit, feedback, no matter how critical, will be very much appreciated (to paul.engel@ucd.ie). Comments and suggestions from nursing or other healthcare students would be especially welcome.

TOPIC 2

Remarkableness of life

What's so special about life?

Can we put our finger on what it is, in a general sense, that is so special about life and living structures? As already mentioned, there are a large number of remarkable processes to describe and explain, but I propose to summarise what all of them have in common by contrasting two very different sorts of object: a sleek new sports car and a living cell (Fig. 2.1). The modern sports car is a triumph of precision engineering and has a large number of gadgets and components to give us a smooth, safe, controlled ride. It is also eye-catching and commands our attention and is physically strong and durable. The cell, by contrast, is tiny and fragile, too small on its own to be noticed except under a microscope.

Self-direction

Let us, however, look at the unequal comparison a little harder. The car may have 0–60 mph acceleration in under 10 s, power-assisted steering, satnav and all the rest, but without someone at the wheel it is going nowhere. The cell on the other hand can rove around on its own initiative, exploring surfaces, changing shape, perhaps engulfing the odd particle. It does not need a driver.

Energy source

What about the fuel requirements? They both need fuel in order to power their activities, but if the car has a petrol engine you will be in trouble if you fill up with diesel (or vice versa). The cell will not thank you for either petrol or diesel, but, on

Pain-Free Biochemistry   Paul C. Engel
© 2009 John Wiley & Sons, Ltd

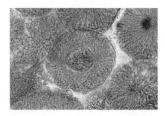

Figure 2.1 The dimensions in the image of the sports car are about 100 times smaller than reality. The mammalian cell, on the other hand, is shown with a diameter 700 times *larger* than reality, as seen under the microscope. (Porsche car image reproduced courtesy of Shutterstock.com)

the other hand, depending a bit on just what sort of cell it is, it is likely that it will do pretty well on any one of a variety of different **sugars**. It might be reasonably happy to keep going with amino acids – fats might be another option; the cell in fact is likely to be remarkably versatile in its ability to use different energy sources. What is more, the cell goes out to find or soak up its fuel and does not depend on someone else to put a nozzle in its tank.

Self-assembly and replication

Just as the car needs a driver, at the assembly stage also it is manufactured according to an external blueprint on an assembly line, with people and other machines to put the parts together. The car itself cannot help the process along in any way, and if the assembly line shuts down, the car remains unfinished. The cell carries its own blueprint, and it interprets and carries out the instructions all on its own. A human cell might need a suggestion or a signal to get going, but once it gets the message it is autonomous. It makes or assembles all the parts it needs, and perhaps the most remarkable thing of all is that one cell will turn into two identical daughter cells as you watch it (Fig. 2.2a). If the Rolls stopped at the traffic lights turns into two Rolls, you should stick to the tomato juice! (Fig. 2.2b).

(a)

(b)

Figure 2.2 (a) The bacterial cell is shown undergoing its normal process of reproduction by binary fission, i.e. splitting into two identical daughter cells. Depending on the type of cell and the growth conditions bacteria can often accomplish this feat every half-hour or so! (b) An unlikely scene of a car undergoing binary fission.

Carrying the blueprint

Small and fragile maybe, but living cells have quite a bit to shout about in terms of their own precision engineering. If you consider what they can do, it is clear that, packed inside these tiny structures, there has to be an enormous amount of information content. The blueprint is somehow contained in the chemical substances that make up the cell, and, if we think about how the blueprint is put into action and about the fact that the whole structure assembles itself unaided, it is difficult to escape the idea that at least some of the biological molecules must be large and complex. Self-assembly implies a precise shape, to guide each piece in the molecular jigsaw puzzle into the right slot beside the right neighbours, and we shall see that shape is indeed critically important.

Thoughts about size

Superficially, there might seem to be a problem here. If cells are so small and their molecules are so large (and there are a lot of different kinds), how can we possibly fit all the molecules in? It sounds like the impossible pile of clothes on the bed beside the suitcase that is miles too small! The answer to this puzzle is that we are really talking about two different and unrelated scales of size: when we say the cells are small we are relating their size to what we humans can or cannot see with the naked eye. When we say that the molecules are large, we are not making any sort of statement about visibility. We are comparing them to much smaller chemical objects, far smaller than can be seen even under the most powerful microscope. Fortunately, this leaves enough space for biochemistry to do what it has to do! Even though in molecular terms these are big molecules, there is plenty of space for lots of them inside the average cell.

CHEMISTRY I

The basic structure of substances: atoms, molecules, elements and compounds

Do you need this section?

This is a reminder of (or introduction to) basic chemistry. If you have recently done and understood school or college chemistry, you can probably skip the whole of the next two chapters, which are here to make sure that everyone is happy with atoms, molecules, **valency** and other basic chemical concepts that we are going to need to use and take for granted as we move on through the biochemistry.

Does biology obey chemistry?

Until well into the nineteenth century, chemists thought that biology occupied a separate world with its own rules, different from those of ordinary chemistry, and involving mystical 'vital forces'. It gradually became obvious, however, that, special though they seem, living things obey exactly the same chemical rules as inanimate Nature does. The substances inside living organisms are made up of the same fundamental building blocks as are used in the world outside. The only difference is that, as we have already mentioned, some of these substances are remarkably complex. Coming together in the full architecture of the cell, they perform the intricate network of reactions that we call 'life'.

Pain-Free Biochemistry   Paul C. Engel
© 2009 John Wiley & Sons, Ltd

The building blocks – atoms, elements and molecules

If you look at a shiny, gold ring it looks smooth and continuous, likewise a glass of water – both the glass and the water. But, even back in ancient Greece, philosophers wondered whether matter, the multitude of solid, liquid and gas substances that make up our world, was truly continuous or might perhaps be made up of tiny particles too small to see. Although they thought deep thoughts, they did not have ways to test their ideas. Humans had to wait patiently for a couple of thousand years for definite answers. At last, the experimental scientists of the seventeenth and eighteenth centuries produced the evidence. They concluded that:

1 Every pure substance (pure gold, pure water, pure oxygen, etc.) is made up of minute particles called **molecules**, all of them identical for a particular substance, so that every water molecule is like all the other water molecules. The water molecules are also different from all the molecules of oxygen or of any other substance.

2 Many substances are mixtures. A lump of granite, a bucket of sea water, a breath of fresh air and a piece of cake are all made of many different substances and therefore contain different kinds of molecules. In the case of the granite you can see separate bits of different substances – in other words, it is not very thoroughly mixed, but there is nothing immediately visibly obvious to tell us that air or clean sea water are mixtures.

3 Although there are vast numbers of different chemical substances, and accordingly vast numbers of different kinds of molecule, they are all made up from a much more limited set of smaller units called **atoms**. In fact, there are only a hundred or so different kinds of atom, each different sort with its own name. In the same way, a few dozen fixed kinds of Lego block will allow you to build an unlimited number of different structures.

4 Substances made up of only one kind of atom, carbon, hydrogen, oxygen, iron and so on, are known as **elements**. Therefore, since there are a hundred or so kinds of atom, there are also a hundred or so elements.

5 Most atoms have to have partners – rather than travelling around as lonely individuals, they form stable couples or families. This is what molecules are – precise groupings of exact numbers of certain kinds of atom linked together. Thus, oxygen atoms generally prefer to exist in pairs. The oxygen atom is given the symbol O and so the oxygen molecule is written as O_2 to indicate the pairing (Fig. I.1). Occasionally, a threesome is formed, and the different **formula**, O_3, makes this a distinct chemical compound, ozone in fact.

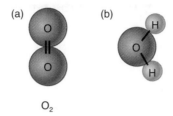

O_2

Figure I.1   (a) Pairing of atoms in an oxygen molecule. (b) Combination of different kinds of atom in a water molecule.

6 The pure substance water has a molecule made up of two different kinds of atom, oxygen and hydrogen, and in this case two hydrogen atoms (H) team up with a single oxygen atom (O) to make H_2O (Fig. I.1). The formula H_2O tells you what kind of atoms are in the molecule and how many of each. In the case of the gas, carbon dioxide, as the name implies (di $= 2$), one atom of carbon (C) combines with two oxygen atoms to give CO_2. Marriages of this sort between two or more different kinds of atom are called **compounds**.

Self-test MCQs on Chemistry I

1 Brass is made by melting together copper and zinc in variable proportions. Brass is therefore

(a) an atom? (b) an element? (c) a molecule? (d) a mixture?

2 CH_4 is the basic unit of the gas methane. Methane is therefore

(a) a compound? (b) a formula? (c) a mixture? (d) a molecule?

3 Hydrogen peroxide (found in hair rinses, washing machines, etc.) has the formula H_2O_2. Which is true?

(a) Hydrogen peroxide is a mixture of hydrogen and oxygen.

(b) Hydrogen peroxide is a mixture of water and oxygen.

(c) Hydrogen peroxide is a compound of hydrogen and oxygen.

(d) Hydrogen peroxide is a molecule of water and oxygen.

CHEMISTRY II
Atomic structure, valency and bonding

Fleas with little fleas: subatomic particles

As we have seen, chemists worked out over time that the world around us is made up of a huge range of different substances, usually mixed up together, but occasionally separate and pure. Each of these substances is made up of molecules, and the enormous number of different kinds of molecule is made up, in turn, of a limited set of possible atoms: carbon atoms, oxygen atoms, hydrogen atoms and so on. Molecules are therefore the smallest possible unit of any substance. However, in seeking to understand the difference between different sorts of atom and also what is actually happening when they combine in molecules, chemists and physicists eventually deduced that atoms themselves are made up of even smaller units called subatomic particles. Amazingly, the hundred plus different kinds of atoms are made up of different combinations of just three kinds of particles! These are

- **protons**, which carry a positive electrical charge;
- **electrons**, which carry an equal and opposite negative charge; and
- **neutrons**, which have no charge.

Every atom of every chemical element, carbon, oxygen, gold and all the others, is made up of combinations of different numbers of just these three kinds of particle.

Pain-Free Biochemistry   Paul C. Engel
© 2009 John Wiley & Sons, Ltd

Elements and atomic number

We can define each element according to how many protons there are in its atom – its **atomic number**. So in Table II.1, the smallest of all, hydrogen, has atomic number 1 because it has only one proton; the carbon atom, quite a bit bigger, has six protons and so has atomic number 6, and so on. The number of positively charged protons is always exactly matched by the same number of negatively charged electrons, so that the charges cancel out and the complete atom is electrically neutral.

What about the neutrons? For small- to medium-sized atoms, the number of neutrons tends to be similar to, although not necessarily equal to, the number of protons; for large atoms, the number of neutrons is quite a bit larger than the number of protons. Examples are shown in Table II.1.

Table II.1 Composition of atoms

Element	Atomic no.	Protons	Electrons	Neutrons
Hydrogen	1	1	1	0
Carbon	6	6	6	6
Iron	26	26	26	30
Uranium	92	92	92	146

The protons and neutrons cluster together in the centre of the atom, forming its **nucleus** (not to be confused with a cell's nucleus!); the electrons orbit around the nucleus (Fig. II.1).

Relative atomic mass

Apart from charge, these sub-atomic particles have one other property that we have not mentioned – weight. The weight of the particles in turn makes up the overall weight of the atoms, and so also the molecules and finally the substances made up of those molecules. We know by experience that matter has weight, and so the tiny particles that make up matter must of course have weight, since there is nothing else

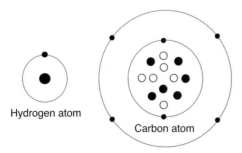

Hydrogen atom

Carbon atom

Figure II.1   Atomic structure of hydrogen and carbon atoms.

there! In fact, the weight is nearly all contributed by the protons and neutrons. The electrons, in spite of their charge, are very much lighter, about 2000 times lighter than a proton. Since the actual weight individually of all these particles, electrons, protons, neutrons, and even the atoms and molecules, is so minute, most of the time it is more useful to think, not of their actual absolute weight, but rather of their relative weights, i.e. their weights compared to one another. We can simply decide to define the weight (chemists prefer to use the word 'mass') of a proton and a neutron each as one unit (without worrying about what tiny fraction of a gram that one unit might be) and an electron as virtually zero. On that basis, from the list above, the relative atomic mass of hydrogen is 1, carbon is 12 (6 protons plus 6 neutrons), iron 56 (26 + 30) and uranium 238 (92 + 146).

Relative molecular mass

If we add oxygen to our list, with its relative atomic mass of 16, we can now go a step further and calculate the relative molecular mass of water, H_2O, as $2 + 16 = 18$ and carbon dioxide, CO_2, as $12 + 32 = 44$. This means, then, that one molecule of carbon dioxide is 44 times as heavy as one atom of hydrogen. Relative molecular mass is given the symbol M_r, so, for CO_2, $M_r = 44$.

Balanced chemical equations

When a chemical reaction occurs, the atoms rearrange into different partnerships, but nothing gets lost. All the atoms that were there at the beginning have to be there, somewhere, at the end. For example, if we burn hydrogen in oxygen we get water. The three molecular formulae for these substances are H_2, O_2 and H_2O, and so we could state our chemical reaction here as:

$$H_2 + O_2 \rightarrow H_2O$$

However, if we do a bit of bookkeeping, there are two atoms of oxygen on the left and only one on the right, implying that matter has disappeared – strictly not allowed! What we have written out is a chemical **equation** but it is clearly wrong: equations must balance so that every atom is accounted for. If we rewrite the equation as follows:

$$2H_2 + O_2 \rightarrow 2H_2O$$

then all is well – four atoms of hydrogen on the left and on the right, two atoms of oxygen on the left and on the right, a balanced equation (Note: the number in front of any chemical molecular formula, like the 2 in front of H_2O, refers to the *whole* molecule; it stands in this case for two molecules of H_2O). Since we also now know the relative molecular masses of these three substances, 2, 32 and 18, this equation also tells us that $2 \times 2 = 4$ g of hydrogen will react with 32 g of oxygen to give $2 \times 18 = 36$ g of water. For that matter, 4 tons of hydrogen would combine with 32 tons of oxygen to give 36 tons of water – with an extremely big bang!

The equation is OK but does it happen?

It is worth noting that, in spite of what we have just said, just mixing the hydrogen and the oxygen will not give water. Most reactions need a little encouragement, and in this case a flame or a spark will do. The heat, a bit like drinks at a party, gives some of the molecules enough energy to rush at one another with the necessary enthusiasm. Their initial reticence is just as well. Lots of reactions that might theoretically be possible just will not happen at an appreciable rate without some help. As we shall see presently, this gives the cell a chance to maintain a bit of order!

Combining ratios

We now know that atoms combine with one another in a strict numerical fashion, and it is certainly not always monogamous pairing. Thus, oxygen takes on *two* hydrogens to form a threesome called water, as we saw above. In CO_2, carbon dioxide, however, a single oxygen is not sufficient to cope with one carbon atom; it takes two oxygen atoms. These are not just random numbers. The principle governing these combining ratios is called valency. Hydrogen has a valency (a combining power) of one, oxygen two and carbon four. Accordingly, if a single carbon atom combines with hydrogen, it requires four hydrogen atoms to satisfy the greedy appetite of the carbon. This makes the molecule of methane, CH_4.

Stable electron shells

How does this work? It all depends on the number of electrons in each atom. The electrons assigned to each atom are not just randomly arranged; they are organised in sets or 'shells', each of which can only hold a fixed maximum number of electrons. As each shell fills up, any new electrons have to move into the next shell. The innermost shell, closest to the nucleus, can only hold a maximum of two electrons – one in hydrogen, two in atom no. 2, helium. This means that for atom no. 3 (lithium), with three electrons, we have to start a new shell a little further out. Beyond the first layer, electron shells can hold eight electrons. In fact, full sets of electrons seem to be a very stable arrangement, one that the average atom aspires to for its own outer shell of electrons. It can achieve this idyllic state by one of three possible strategies:

- It can grab extra electrons from another atom to fill up an incomplete shell.

- It could give away a few unwanted electrons from its outer shell to expose a complete shell underneath.

- It might come to a civilised sharing arrangement with another atom.

Covalent bonding

Consider the water molecule again. Oxygen has six electrons in its outer shell, too many to give away lightly. On the other hand, it only needs another two to complete a set of eight. Each of the hydrogens has only one electron. Remember, however, that they are special, having only one shell, one that would be complete with just two electrons. So they each need only one more. If each of the hydrogens shares its one electron with oxygen in exchange for a share of one of the oxygen's electrons, then everyone is happy, oxygen with eight and the hydrogens with two each. This kind of sharing arrangement between two atoms is called a **covalent** bond (Fig. II.2).

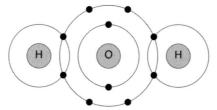

Water molecule showing shared electrons

Figure II.2 Electronic structure of a water molecule. Reproduced from Tortora and Derrickson (2009) *Principles of Anatomy and Physiology*, 12th edn, Wiley International Student Version, New York. © 2009 John Wiley & Sons. Reprinted with permission of John Wiley & Sons, Inc.

Ionic bonding

Now let us look at two more types of atom, two we have not mentioned till now: sodium and chlorine. In passing we must note that, while hydrogen (H), oxygen (O) and carbon (C) all have the luxury of their first letter as a symbol, 26 letters are not enough to go round 100 elements. In the case of chlorine, since C is already spoken for, the symbol is Cl. For sodium there is an additional problem: elements are sometimes called different names in different languages and sodium is known as natrium in German-speaking parts of Europe, so that instead of fighting it out with sulphur for letter 'S', sodium has to contend with nitrogen for 'N'. Nitrogen gets 'N' and so sodium becomes Na. The symbol for sodium chloride is not SC, therefore, but NaCl, indicating that these two types of atom join up in 1:1 pairs. These two, however, do not share electrons. Sodium has only one electron in its outer shell and chlorine has seven. If sodium went into a pair-sharing arrangement such as we have just seen, chlorine would achieve its eight, but sodium would gain only one, taking it up to two when it also needs eight. If instead sodium donates its electron completely, then chlorine has a full complement, but so does sodium, because its next layer has a complete set of eight (Fig. II.3). This kind of link between two

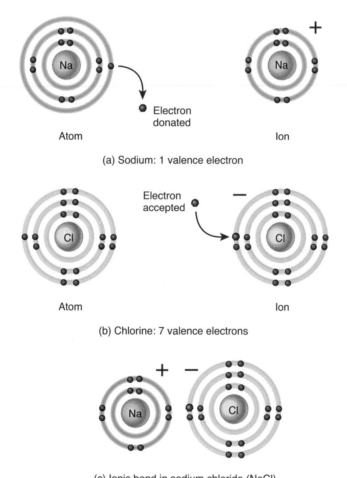

Atom   Ion

(a) Sodium: 1 valence electron

Atom   Ion

(b) Chlorine: 7 valence electrons

(c) Ionic bond in sodium chloride (NaCl)

Figure II.3 Donation and gain of an electron to form an ionic bond in sodium chloride. Reproduced from Tortora and Derrickson (2009) *Principles of Anatomy and Physiology*, 12th edn, Wiley International Student Version, New York. © 2009 John Wiley & Sons. Reprinted with permission of John Wiley & Sons, Inc.

atoms is called an ionic bond. When two atoms pair up through an ionic bond, the compound they make is called a **salt**. It just so happens that this particular salt, sodium chloride, is the one that we all refer to as 'salt', i.e. table salt, but chemically speaking potassium bromide, lithium iodide, calcium chloride, magnesium fluoride, etc. are all salts, even though you might not want them on your fish and chips!

The magic of carbon

Returning then once more to carbon, we have seen that carbon has six protons and six electrons. Six electrons give a full shell of two and an incomplete shell of four. This accounts for the valency of four because carbon can make up its eight by sharing four electrons from other atoms, e.g. from four Hs in methane or from two Os in carbon dioxide. The four valency 'arms' are entirely equivalent to one another, and accordingly they are arranged evenly and symmetrically around the C atom. Quite often this leads to a molecule like methane being drawn as a cross; a cross is symmetrical on our two-dimensional (2D) pages of print, but chemical atoms are three-dimensional and do not need to have their valencies lying flat on the page! Symmetry in three dimensions means that the four hydrogens in methane will be at the four corners of a tetrahedron (Fig. II.4).

   The simple fact that C has a valency of four is possibly the most important single chemical fact in making life possible. Because carbon atoms have four 'arms', they can each 'hold hands' with a couple of other carbon atoms and still have combining power left over. This means that carbon atoms can join up to form large and intricate structures – such as we find in life (Fig. II.5), but such also as are made synthetically by chemists, e.g. plastics and fibres, dyes, drugs, etc. Even outside of biochemistry, the implications of this are so great that the chemistry of this one element, carbon, makes up one major division of chemistry, known for historical reasons as 'organic chemistry', with the chemistry of everything else lumped together as 'inorganic chemistry'. Ironically, in contemporary everyday parlance 'organic' has come to mean just about anything that has escaped the attentions of organic chemistry!

Figure II.4 In a molecule like methane, CH_4, the four valency arms of the carbon atom point symmetrically to the four corners of a tetrahedron. This is usually not shown in 2-D representations of chemical structures, but, to emphasise the point, the bonds can be drawn so that the ones coming towards you (out of the page) seem to be getting thicker, and the ones going away into the page taper into the distance, getting smaller.

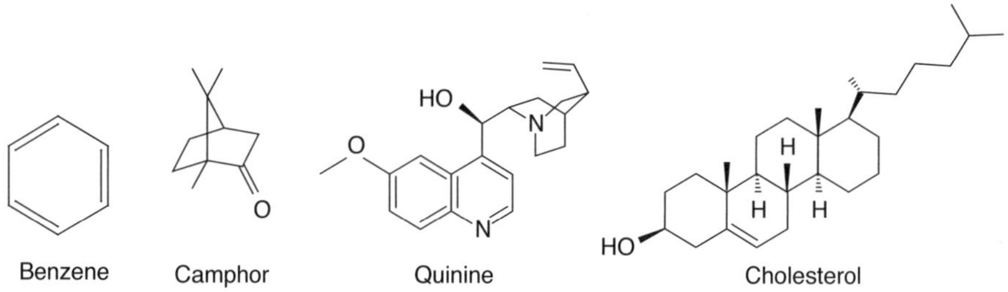

Benzene   Camphor   Quinine   Cholesterol

Figure II.5 Complex bio-organic molecules.

Chemical shorthand

Looking at the structures in Fig. II.5, if you have not done much chemistry be-fore, you may be mystified and may wonder where the atoms are! This is because organic chemists have adopted a shorthand notation to save time, space and ink! Since biochemicals are all organic chemicals, based on chains and rings of carbon atoms, biochemists too use the same notation. If you see a zig-zag line with no atoms marked on it or a hexagon similarly with no atoms marked, you assume that each corner in the structure is a carbon atom. The second shortcut becomes obvious if you think about one of these invisible C atoms in the middle of a zig-zag line; the zig-zag line tells you that the carbon atom is 'holding hands' with two neigh-bouring C atoms, but this only accounts for two of the four valencies. What about the other two? The second shortcut is that if the other partners are 'only' hydrogen atoms they are not written in; you just assume that they are there. In the shorthand

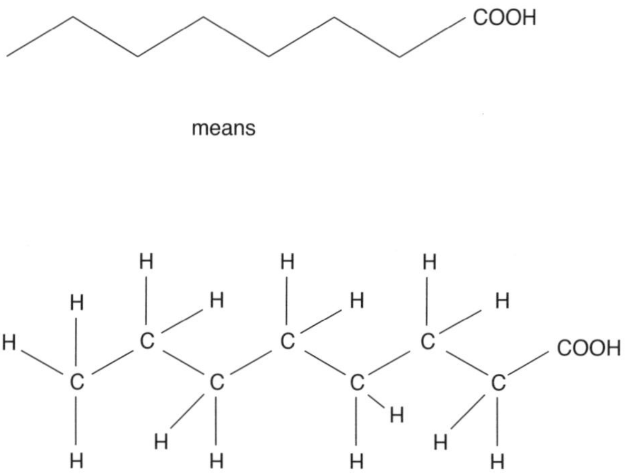

COOH

means

Figure II.6 Octanoic acid. The two structures show exactly the same molecule. In one case every atom is explicitly shown; in the other the chemists' shorthand assumes that it is obvious that there is a carbon at every corner and that if nothing else is shown all the spare positions are filled up with hydrogen atoms.

structure for benzene in Fig. II.5 it looks as if each of the (invisible) carbon atoms at the corners of the hexagon has only three valencies satisfied. In fact, each of them has an H atom attached as well as its links to the other carbon atoms; it simply is not shown. Anything else among the carbon atoms or hanging off them – oxygen, nitrogen, sulphur – is always explicitly written. You will see more of these structures as you go through the book, and anything that is not explicitly shown by its atomic symbol will be carbon or hydrogen, e.g. Fig. II.6.

Self-test MCQs on Chemistry II

1 Carbon monoxide (highly poisonous, unlike carbon dioxide) has the formula CO.

What is its relative molecular mass?

(a) Atomic no. 28

(b) 28 protons

(c) 14 protons and 14 neutrons

(d) 28

(e) 28 g

(f) 12:16

2 Methane can be burnt to give carbon dioxide and water. Which of the following is the correct equation for this reaction?

(a) $2CH_4 + 5O_2 \rightarrow 8H_2O + 2CO$

(b) $CH_4 + 3O_2 \rightarrow CO_2 + 4H_2O$

(c) $CH_4 + 2O_2 \rightarrow CO_2 + 2H_2O$

(d) $CH_4 + O_2 \rightarrow CO_2 + H_2O$

3 Lithium, the lightest of all the metal elements, has only three electrons in total; chlorine has an outer shell with seven electrons. If these two elements pair up, what is the most likely way for this to happen?

(a) Lithium donates its three electrons to the chlorine to make ten.

(b) Lithium receives seven electrons from chlorine to give an outer shell of eight for both atoms.

(c) Lithium donates one electron leaving a stable innermost shell of two and giving chlorine eight in its outermost shell.

(d) The two atoms share their eight outer-shell electrons so that each has four.

CHEMISTRY III

Protons, acids, bases, concentration and the pH scale

See also Appendix 1.

Ions in solution

In Chemistry II, we saw that two atoms such as Na and Cl can establish an ionic bond by, respectively, fully shedding and fully gaining an electron. In this way, both atoms achieve stable outer shells of eight electrons. In doing so, both atoms also acquire a net electrical charge because the numbers of protons and electrons no longer balance. In this state they are called **ions**. Chlorine, having gained an electron, is now written as Cl^- to indicate the single net negative charge of the chloride ion, and the sodium ion, now with more protons than electrons, has a net positive charge and so is written as Na^+. Positive and negative objects tend to attract one another but, unlike the situation with a covalent bond, the association in an ionic bond is not like marriage for life. In a solution of sodium chloride, the individual ions are not tied to their original partners; they all float freely as separate independent ions. All that is required is that, like at a dance, the number of positive ions matches the number of negative ions.

The elements that most easily form positive ions are usually the metal elements – iron, copper, calcium, etc. as well as sodium – but another element that can also easily form a positive ion is hydrogen. We have seen that hydrogen can form a complete shell of two electrons by gaining a share of an electron from another

Pain-Free Biochemistry   Paul C. Engel
© 2009 John Wiley & Sons, Ltd

atom, as it does in forming a water molecule or a methane molecule. Potentially, however, it could jump the other way and escape the situation of an incomplete shell by losing its one electron rather than gaining a second one. If it does this, all that is left is the single naked proton. This, as we know, is positively charged and can be written as H^+, since it is in fact a hydrogen ion.

Losing protons: acids

In biochemistry, there are many compounds where the hydrogen atom seems unable to decide whether to share in stable partnership or to abandon the electron and strike out on its own as a hydrogen ion (H^+). This depends not only on the whim of the hydrogen atom but also on the greed of its partner atom for the electron. Chlorine, as we have seen, is eager to acquire sole use of the electron from sodium and the same is true if we combine H and Cl. Most of the time HCl molecules split up into ions, H^+ and Cl^-, and because of this tendency to shed H^+ very readily, HCl is said to be very acidic. Acids are compounds that readily lose a proton, H^+. HCl, then, is called hydrochloric acid, found on the shelves of chemistry laboratories, but also found in your stomach to give it its digestive acidity.

HCl is regarded as a strong acid because it loses H^+ with such ease. The chloride ion can only be persuaded to pick up an H^+ again if there are a lot of them around so that the Cl^- is literally bombarded with suitors. There are, however, many weaker acids in biochemistry that hang onto their H a bit tighter. There is a group of atoms that occurs over and over again in many of these biochemical acids. This is the **carboxyl** group —COOH, and one of the simplest compounds (carboxylic acids) that has this structure is CH_3COOH (Fig. III.1).

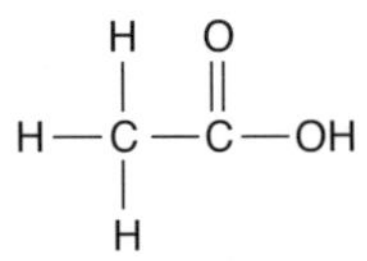

$$
\begin{array}{ccc}
& H & O \\
& | & \| \\
H - & C - & C - OH \\
& | & \\
& H &
\end{array}
$$

Figure III.1 Acetic acid (also sometimes called ethanoic acid).

Chemical 'groups'

Note that the formulation we have just used for our carboxylic acids is another bit of chemists' shorthand. It seems to suggest that the three H atoms are joined to both carbons and that the two oxygen atoms are joined to one another. Neither is true, as we see in the figure. CH_3 is an accepted shorthand for a recurrent group of atoms (the **methyl** group), a cluster of combined atoms that crops up again and again in different compounds, likewise —COOH (see also Fig. II.7). Occasionally, to avoid ambiguity, chemists might put a dot between the groups CH_3.COOH.

Acid and ion

The compound CH_3COOH, acetic acid, is the main component of vinegar (apart from water). Since it is an acid it must be able to lose a proton. Sure enough, the —COOH (carboxylic) group is capable of becoming the negatively charged —COO$^-$ (carboxylate ion) with the loss of a proton, but it does this considerably more reluctantly than HCl does, which is why it is regarded as a weak acid. Nevertheless, under physiological conditions most of the biochemical acids we shall encounter exist almost entirely in the deprotonated state as the ion. Note that we do not think of the protonated carboxylic acid (e.g. acetic acid) and the deprotonated carboxylate ion (e.g. acetate) as different compounds, but just as different, readily interchangeable forms of the same compound. This has led biochemists to be careless in talking about these compounds, using the names of the acid and the salt interchangeably, e.g. a lecturer (or a book) might talk about production of 'lactic acid' (see Topic 13) and then in the next breath about utilisation of 'lactate'; *both* terms are intended to refer to the same ionisable acid that will in fact be present overwhelmingly as lactate ion under physiological conditions.

Gaining protons: bases

A neutral substance that can lose a proton to form a negative ion is an acid, but there are also neutral substances that do the reverse and pick up a proton to form a positive ion. These are called **bases**. An example in biochemistry is the so-called **amino group**, which has the structure —NH$_2$. This indicates that the nitrogen atom N has two hydrogen atoms attached but the dash tells you that it still has one unsatisfied valency. So it is not a complete compound; it needs to join onto something else. It could, for example, join onto a methyl group (—CH$_3$). In chemists' shorthand this is written as CH_3NH_2. Again, this formula seems to imply that the H_3 and the N are directly linked, but they are not. It is understood that the bond linking the two 'groups', methyl and amino, is between the C and the N (Fig. III.2). As above, this can be made more explicitly clear by writing CH_3.NH$_2$ but a chemist would not normally do so, understanding the structure without the need for a dot. For our present purpose, the important thing is that the amino group, —NH$_2$, readily picks up a proton to become —NH$_3^+$. Just as we saw that acids have a strong tendency to lose a proton, compounds that are 'very basic' or 'strong bases' have

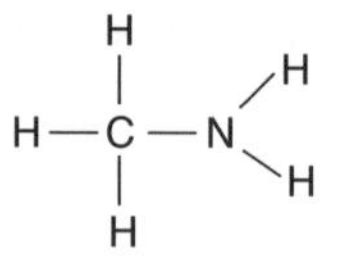

Figure III.2　Methylamine.

a very strong tendency to do the opposite and to be in the 'protonated' state, although once again this will depend also on how many protons there are around. The more there are, the bigger the chance that one will have hopped on at any given moment.

Concentrations of substances

Thinking about how many protons there are around makes this an appropriate point to introduce the all-important concept of **concentration**. If we want to say something meaningful about the chances of bumping into a proton – or a molecule of oxygen or glucose or anything else – we need to say how much of that substance there is in a given space. There are basically two ways of doing this. The first is simply concentration w/v, which stands for **w**eight per unit **v**olume. A concentration defined in this way might be 10 g/l or 0.5 mg/ml or 2 mg/dl. The last one, a concentration per decilitre (0.1 l or 100 ml) would be quite often encountered in nursing. As for the weights, the metric system tends to work in factors of 1000, and in biological/medical situations one is usually working in grams or less. Thus, milligrams (mg) are 1/1000 g and micrograms (μg) are 1/1 000 000 g (the symbol in front of the g is the Greek letter 'mu', μ, used for 'micro'). Occasionally in dealing with potent substances like hormones, you might even get a thousandfold lower, to nanograms (ng).

The 'mole'

Why do we need any other way of expressing concentration? A statement of concentration w/v is a good recipe for preparing a solution, but it does not tell us anything directly about how many molecules of the substance we have dissolved. Since we know that chemical reactions involve exact numbers of molecules of each compound reacting in a precise ratio, it would be helpful to build that information in somehow. This is done by defining the **molar** concentration (Table III.1). This has nothing to do with teeth. It is the number of 'moles' of a substance per litre. But what is a **mole** in this context? We have already met the idea of relative molecular

Table III.1   Different ways of expressing concentration

Type of concentration	Typical units	Equivalence
Weight/volume (w/v)	g/l (grams per litre)	1 g/l
	mg/ml (milligrams per millilitre)	= 1 mg/ml
	mg/dl (milligrams per decilitre)	= 100 mg/dl
Molar concentration (molarity)	M (molar = moles per litre)	1 M
	mM (millimolar = millimoles per litre)	= 1000 mM
	μM (micromolar = micromoles per litre)	= 1000 000 μM

mass – 18 for H_2O, 32 for O_2 and so on. We now simply (and arbitrarily) define the mole as being 18 g for water, 32 g for oxygen, etc. so that:

1 mole of any substance is a number of grams equal to its relative molecular mass

These relative masses are useful in practical chemistry and biochemistry, because they define the proportions in which substances react.

Molar concentrations

Earlier we noted that 4 g hydrogen will react with 32 g oxygen to give 36 g water. We could now rewrite these amounts another way and state that two moles of hydrogen will react with one mole of oxygen to give two moles of water. Not surprisingly, this reflects the original chemical equation for the reaction. It means that if we want to talk about concentrations in a way that is relevant to their chemical combining/reacting potential, it is more helpful to talk in moles per litre than gram per litre. Mole per litre is given the term molar and the symbol M. Thus, for example, 2 M NaCl means two molar sodium chloride, containing two moles of sodium chloride (approximately 116 g) per litre.

Just like units of weight and volume, the scale of molar concentration goes down in metric factors of 1000, and in biochemistry substances are generally in the millimolar (mM), micromolar (μM) or even nanomolar (nM) ranges of concentration.

The pH scale

Finally, let us return briefly to acids, bases and hydrogen ions. Under most physiological situations, the molar concentration of the hydrogen ion (H^+) in solution is close to 0.1 μM, which we can also write as 10^{-7} M. This is also described as 'neutral' solution. We can move away from neutrality by either increasing the H^+

concentration, by adding an acid, or decreasing it by adding a base, which will trap most of the protons by combining with them. A solution at, say, 10^{-3} M H^+ would be described as acidic and one at, say, 10^{-9} M would be said to be basic or alkaline. For convenience, a shorthand terminology is generally used to describe these wide ranges of H^+ concentration, the pH scale. You will meet it constantly and need to be familiar with it. Thus, a solution with 10^{-7} M H^+ is said to be at pH 7 (Note: 7 and not -7), one at 10^{-3} M H^+ is at pH 3 and so on. So a high pH like 10 or 11 is very basic (alkaline), a low one like 2 or 3 denotes very acid conditions and, as we have seen, pH 7 is neutral (see Appendix 1).

Self-test MCQs on Chemistry III

1 Methyl alcohol (ethanol) has the formula CH_3OH. What would you expect the actual linkages between atoms to be?

(a) H–H–H–C–O–H

(b) C–H–H–H–O–H

(c) 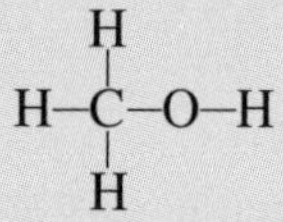
$$\begin{array}{c} H \\ | \\ H-C-O-H \\ | \\ H \end{array}$$

(d) 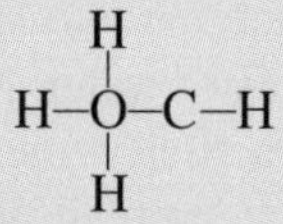
$$\begin{array}{c} H \\ | \\ H-O-C-H \\ | \\ H \end{array}$$

2 18 g glucose is dissolved in 0.2 l water. What is the concentration in mg/dl?

(a) 9

(b) 36

(c) 0.036

(d) 9000

3 The relative molecular mass of glucose is 180. What is the molar concentration of the glucose solution in question 2?

(a) 0.5 M

(b) 0.2 M

(c) 2 M

(d) 0.05 M

4 Solution A is at pH 4. Solution B is at pH 2. The H^+ concentration in A is which of the following?

(a) 100 times that in B

(b) 2 times that in B

(c) 100 times less than in B

(d) half that in B

5 Water, H_2O, has a relative molecular mass of 18. What is the molar concentration of water itself?

(a) 18 moles per litre

(b) 18 g/l

(c) 18

(d) 18 M

(e) 55.5 litres per mole

(f) 55.5 M

(g) 55.5

TOPIC 3

Shape, molecular recognition and proteins: an example

See also Appendix 2.

Immunity

How is chemistry going to help us understand the processes of life? Let us take one specific medical example – immunisation. If we are exposed to 'foreign' substances such as chickenpox virus, incompatible red blood cells in a blood transfusion or TB (tuberculosis) bacteria, it wakes up molecular sentries and 'hitmen', which are able to respond immediately in the event of a second invasion. These sentries are totally committed, in the sense that there is a different sentry or set of sentries detailed for each potential attacker – vaccination against smallpox does not protect against cholera; exposure to TB will not help you resist malaria. But how does this security system work night and day without our being remotely aware of it? There is no little man behind a desk, no computer, no bar code, no CCTV. This is chemical surveillance, and it requires very precise molecular recognition. What kind of molecule can do this? It needs a large molecule and one with an exact shape that can match a complex shape on the surface of the intruding virus or bacterium or parasite. It needs to be a complex shape, as otherwise there would constantly be false alarms. Just as in wartime you do not want to shoot down your own aircraft or torpedo your own submarines, so also the defence system of the body needs to be very certain about its identification before sending out the 'destroy' message. These tasks of precise recognition in living organisms are generally fulfilled

Pain-Free Biochemistry   Paul C. Engel
© 2009 John Wiley & Sons, Ltd

by proteins. In this particular instance, the group of proteins involved are called 'antibodies'.

Protein, proteins and antibodies

Protein is often talked about in the context of food as if it were a single (good) substance – just as fat is equally misleadingly thought of as a single (bad) substance. In fact, 'protein' refers to a huge family of molecules, tens of thousands of different kinds in humans, of very varied size, shape and function. Although they are all proteins, each kind of protein is a different chemical substance with its own unique molecule. What they have in common (and what qualifies them all as proteins) is that they are all made up of the same set of characteristic building blocks (explored in Topic 4). We have already said that, in order to provide precise shape recognition of a wide variety of different foreign objects, **antibody** molecules must be large. How large? The main serum antibodies, immunoglobulin G (**IgG**), have relative molecular masses (M_r) of about 150 000. Bearing in mind that the M_r of a water molecule is 18, this means that an IgG molecule is more than 8000 times heavier than a water molecule. Another type of antibody molecule, IgM, is six times heavier even than this. So as molecules go, these are indeed very large.

How big are molecules?

On the other hand, exactly how large is a water molecule? Our figure of 18 for its M_r has no units. In Chemistry I, we gave everything comparative weights, but we never specified in absolute terms, i.e. in grams, how much a single proton or a hydrogen atom or a water molecule actually weighs. We managed to avoid the question by introducing the idea of moles, so we know that one mole of water weighs 18 g. But how many molecules of water are there in a mole? The answer is a truly enormous number (called **Avogadro's number**), approximately 6×10^{23}. Even if we put this into words as six hundred thousand million million million, it is impossible to grasp. In an attempt to put it across, the famous American chemist Linus Pauling imagined the state of Texas, 262 000 square miles, covered with a layer of fine sand 15 m deep. Each grain of sand is quarter of a millimetre across. This vast amount of sand contains a number of grains roughly equal to Avogadro's number, and that is how many molecules of water there are in a mole (18 g or 18 ml, perhaps quarter of a tumblerful). So the weight of one molecule of water is 18 g divided by that vast number, and the weight of an IgG molecule is 150 000 g divided by the same number. As we said, this is much heavier than the water molecule but is still only 2.5×10^{-19} g – not very heavy!

So again we have the paradox of something that is both very big in one sense and very small in another! Table 3.1 gives a rough idea of the relative sizes (diameters)

Table 3.1   Dimensions of atoms, molecules, cells

Object	Approximate diameter (nm)
Oxygen atom	0.28
Small protein molecule (myoglobin)	6.4
E. coli cell	2000
Typical human cell	50 000

The units of length in this table, nanometres (nm), are 10^{-9} m, i.e. 10 million times smaller than a centimetre. The dimensions given for a human cell are very approximate since human cells vary greatly in shape and size.

of a small atom, a protein molecule and a cell. In thinking about those numbers, you need to bear in mind that the actual volume occupied is proportional to the diameter cubed and also that even the animal cell is too small to see without a microscope.

Molecular recognition by proteins

Fig. 3.1 shows an example of the way an antibody molecule recognises and fits its target, and gives us our first glimpse of the complexity of the shape of a protein molecule. In view of what we have just said about the actual size of these molecules, you might very well wonder how we can possibly know the shape of something so small. The answer lies in a powerful technique called X-ray crystallography (Appendix 2).

Once we have the possibility of complex and specific shapes, all we have to allow ourselves is individual protein shapes that will recognise and 'dock' with the chickenpox virus, red cells of the wrong blood group, and so on. One of the analogies used early on in the study of proteins was that of a lock and key, but there is more to it than a passive fit of two rigid complementary shapes. Many protein molecules can hinge, but also they can undergo more subtle changes, like a glove stretching when it receives a hand.

The recognition by an antibody of its target is a 3-D process, and the wrong hand or a foot or an elbow simply will not give a proper fit to the 'glove', even if it can get in. The extreme **specificity** of antibodies has many consequences and also applications in medicine (Box 1). As we shall see, this principle does not apply only to antibodies. The ability of protein shapes to recognise only the right molecule is used over and over again in enzymes, hormone **receptors**, molecular transporters, structural proteins, etc. Next, however, we need to focus more closely on what a protein is in chemical terms.

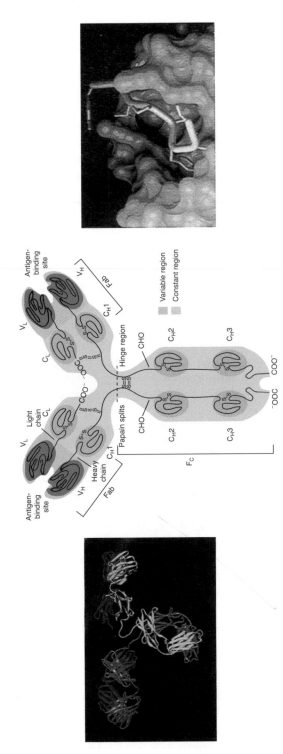

Figure 3.1 The antibody protein molecule fits the structure of its target, the antigen. The left-hand image and the schematic diagram in the middle show the overall structure of a typical immunoglobulin G molecule. The detailed differences that allow each of thousands of antibodies to recognise different structures are at the tips of the two arms of the Y-shaped molecule, which thus has two antigen-binding sites. The right-hand image shows a close-up of the shape of one such site. In this case the antigen is a peptide molecule, shown in pale blue fitting snugly into the shaped cleft in the protein surface. Reproduced from Voet, Voet and Pratt (1999) *Fundamentals of Biochemistry*, 2nd edn, John Wiley, New York. © 1999 John Wiley & Sons. Reprinted with permission of John Wiley & Sons, Inc.

Box 1 Antibodies in medicine

1 **Immunisation.** The recognition of immunity, long before there could be any proper understanding of its basis, led to procedures for deliberate immunisation by exposure to antigens. A famous example is Edward Jenner's experiment on using cowpox to confer protection against smallpox. The experiment would horrify any modern-day ethics committee, but it worked! Immunisation procedures today typically use (a) a weaker live strain of the pathogen, (b) an inactivated version of the full-strength pathogen and (c) an immunogenic fraction isolated from the pathogen. Option (b) is very effective but runs the risk of occasional accidents as happened in the early days of immunisation against polio, when a batch of vaccine for children in Chicago was not, after all, fully inactivated. Systematic, well-managed international immunisation programmes have now eradicated once-widespread scourges such as smallpox and polio.

2 **Inappropriate immune responses.** In the early stages of a child's development, its immune system 'learns' what is local and familiar and presumably unthreatening. It learns, in other words, to distinguish 'self' from 'non-self'. Occasionally, this goes wrong and our immune system starts to attack a component of our own body. This **auto-immune** process is thought to account in large part for various diseases such as multiple sclerosis, rheumatoid arthritis, lupus erythematosus, etc.

Immunity problems led to disastrous results in early attempts at blood transfusion and likewise restricted the timescale over which diabetics could be treated with pig insulin, differing very slightly from human insulin, and therefore recognisable by the immune system as non-self. Immune responses also account for **Rhesus factor** problems in pregnancy, normally brought about when, in a first pregnancy, a mother encounters the blood of her own baby which is carrying an incompatible blood factor. This sensitises the mother, just like a vaccination, so that when the situation recurs in a second pregnancy her immune system rejects the foetus. Immunological rejection is also a big issue in transplantation surgery, which is dealt with, as far as possible, by tissue matching but also by the use of immunosuppressive drugs.

The other side of this coin is seen in the AIDS pandemic of the past 25 years. The virus responsible, HIV, invades and destroys the cells responsible for raising an antibody response and in so doing lays the patient wide open to any opportunistic infection.

3 **Antibodies as tools and drugs.** It is possible to raise antibodies in experimental animals by inoculating with the target substance (several doses over a

few weeks) and periodically bleeding them and collecting the serum. This has opened the door to very sensitive methods of measuring very low concentrations of key substances such as hormones by radioimmunoassay (see Topic 45) or 'ELISA' techniques. Antibodies against cancer cell surface antigens are now also being used to either to deliver potent cytotoxic drugs direct to the intended destination or else as drugs in their own right. This has required that the antibodies be 'humanised' through genetic engineering, since antibodies raised in a sheep or a chicken would inevitably themselves be immunogenic in humans.

Self-test MCQs on Topic 3

1 TB immunity depends on which of the following?

(a) Antibodies produced by the TB organism against our own bodies.

(b) Antibodies produced by our bodies to neutralise the TB organism's antibodies.

(c) Antibodies produced by our bodies in advance in anticipation of possible infection.

(d) Antibodies produced by our bodies in response to the TB organism's proteins, etc.

2 Which of the following statements are untrue?

(i) An antibody molecule is larger than a bacterial cell.

(ii) A bacterial cell is larger than a human cell.

(iii) An antibody molecule is larger than a water molecule.

(a) (i) and (ii)

(b) (i) only

(c) (ii) and (iii)

(d) All three.

(e) (ii) only

TOPIC 4

Proteins: molecular necklaces

See also Appendix 3.

Proteins as chains of amino acids

Although an antibody protein, like every other working protein, has a complex 3D shape of its very own, it is actually made up of a long linear chain of units called **amino acids**, joined end to end like railway carriages. What is an amino acid? Each amino acid molecule has a core carbon atom and three of its four valencies are occupied by (1) an amino group, $-NH_2$, (2) a carboxyl group, $-COOH$, which makes it an acid, and (3) a hydrogen atom. This is what different amino acids have in common. What distinguishes them is what is attached at the fourth position, the so-called side-chain. There are 20 different kinds, like 20 different colours and types of railway carriage. In this case, the train is likely to be several hundred carriages long, so every type is likely to occur several times, although some are much more common than others. The crucial thing is that in every molecule of a particular protein the carriages (the amino acids) will be in exactly the same sequence, even though this looks at first sight quite random. This linear sequence is known as the **primary structure** of the protein. It is not random but follows a very explicit blueprint! We shall see in Topic 35 that the cell's assembly line turns out new protein molecules essentially as linear chains, but one of the remarkable things about proteins is that they appear to know how to fold themselves up (Appendix 3 and Box 2).

Pain-Free Biochemistry   Paul C. Engel
© 2009 John Wiley & Sons, Ltd

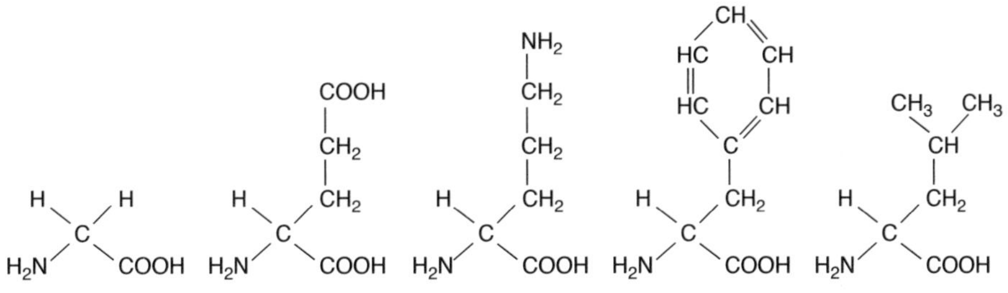

Figure 4.1   Some of the 20 protein-forming amino acids.

The different kinds of amino acid

The 20 side-chains are like a chemical menu, offering a range of sizes, shapes and chemical properties. Just to give you an indication of this, Fig. 4.1 shows a selection of 5 of the 20. The smallest, glycine, has just another hydrogen atom for its 'side-chain'. The next has a side-chain three carbons long, $-CH_2CH_2COOH$, and since the last carbon atom is in a carboxyl group, this is an acidic side-chain, likely to be negatively charged at physiological pH, and the amino acid is called glutamic acid. The third amino acid, lysine, on the other hand, has an amino group at the end of its side-chain, which is therefore basic and will be positively charged at physiological pH. The fourth one is phenylalanine (Phe), and the 'phenyl' label refers to the hexagonal (benzene) ring of carbon atoms. This structure is also described by chemists as an 'aromatic ring' (see also Fig. II.6). Note that half of the linkages in the ring are double bonds – these pairs of carbon atoms are sharing not just one pair of electrons for a single covalent bond, but two pairs. Finally, the fifth amino acid, leucine, has a branched side-chain that is pure hydrocarbon – no oxygens, nitrogens, etc. and no charge.

Joining the amino acids

Note that glutamic acid is called an acid and the others are not, even though they are all amino *acids*! Why is this? We spoke of joining railway carriages end to end, but what do we really mean in chemical terms? If you bring two glycine molecules together, you could envisage joining them by removing $-OH$ from the carboxyl group of one and H from the amino group of the other and linking the $-CO$ to the $-NH$ (Fig. 4.2). This would also make a water molecule, which would disappear among all the other water molecules in the cell. Although this is *not* literally how it happens in the cell, it defines the linkage for us, a linkage known as the **peptide**

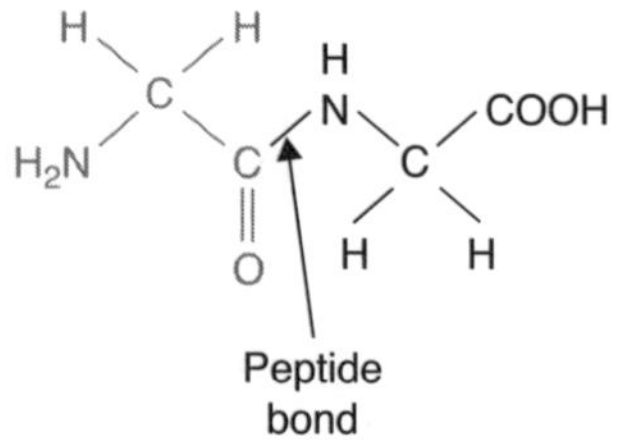

Figure 4.2 A peptide bond between two amino acid (glycine) units.

bond. So if we now imagine a protein molecule 200 amino acid units long, those units will be linked by 199 peptide bonds, forming a so-called **polypeptide** chain. This means that even though all the 200 amino acids are indeed acids, only the very end one will have its —COOH group free because all the 199 others are involved in peptide bonds and unable to lose H^+ to become negatively charged. The side-chain —COOH of glutamic acid, however, is not involved in peptide bonds and so keeps its charge.

Net charge on a protein

Typical proteins, however, do carry a number of negative and also positive charges, mainly contributed by the various side-chains of amino acid units such as glutamic acid and lysine. These side-chains, as we have just said, are not directly involved in forming peptide bonds and so are free to carry a charge. Depending on the precise composition, therefore, a protein will have a characteristic net charge, i.e. if it has 17 negative charges and 14 positive, it will have a net negative charge of 3. This is useful for analysing and separating proteins.

Shape

Under the influence of the various interactions listed in Appendix 3 the long molecular necklace, the primary structure of the protein, folds up. First of all it forms a regular **secondary structure**, but ultimately it goes a stage further to arrange the whole assembly in three dimensions, forming the well-defined shape of an individual protein, its **tertiary structure**, which allows the very specific molecular recognition properties we have already discussed. Finally, many protein molecules are made up not just of a single protein 'blob' but of several, sometimes all identical, like the pieces of a blackberry. This fourth level of structure, if it happens, is called **quaternary structure**.

Box 2 Protein misfolding and disease

The combination of various forces (detailed in Appendix 3), in a balance that depends on the precise amino acid sequence of each protein, leads to a 'folding pathway' and a final, stable folded tertiary structure. Until a few years ago one might have said a final, *unique*, folded structure. However, the understanding of the 'spongiform encephalopathies' – scrapie in sheep, BSE or 'mad cow disease' and related conditions, such as Creutzfeld–Jacob disease (CJD), in humans – has led us to realise that this is not strictly true, at least not for some proteins. The infectious agent in these conditions is the prion protein, and its distinctive feature is that it has two very different folded structures, one that is a normal physiological protein and another that is like a molecular subversive, recruiting its fellow molecules to abandon their correct behaviour and adopt the new way! We have similarly come to recognise that there are a number of other normal proteins that under certain circumstances can go down the wrong folding route and end up as insoluble 'amyloid' deposits. This happens in the brains of Alzheimer's patients, but other proteins behaving in a similar manner lead to complications, for example in long-term dialysis treatment for kidney failure.

Another situation that can lead to misfolding arises when there is a mistake in the amino acid sequence arising from a genetic change (see Topic 36). The protein haemoglobin carries the oxygen round our bodies, and it makes up a large proportion of the total contents of our red blood cells (and accounts for the fact that red blood cells are red!). Its molecule has a quaternary structure (see above), i.e. it is made up of two different kinds of 'blob' or subunit. There are two α subunits, each 141 amino acids long, and two β subunits, each 146 amino acids long. There is a mutation, very common in Africa, that changes just one of these amino acids, at position 6 in each of the two β subunits. Instead of the negatively charged glutamic acid residue that should be at position 6, there is valine, with no charge. As a result of this tiny change in a large molecule, the haemoglobin molecules are able to stack up on one another (only, incidentally, when they have shed their oxygen) and as a result they come out of solution and form long fibres inside the cells. This in turn alters the whole shape of the red cell, which takes on a long, rigid shape, quite different from the normal flexible disc shape. This gives the condition its name: sickle cell anaemia. The 'sickling' tends to take place in the capillaries, thus obstructing blood flow, and it also leads to haemolysis, bursting of the erythrocytes. Through immigration, this condition is quite frequently encountered in European clinics, and likewise is common in Afro-American populations. Thus, a very small and subtle change in just one protein can cause serious disease.

Self-test MCQs on Topic 4

1 If a peptide is made up of glutamic acid (at the amino end of the peptide) joined to glycine joined to lysine joined to glutamic acid (at the carboxyl end of the peptide), which of the following correctly describes the charges on the peptide at neutral pH?

   (a) 2 negative 1 positive, net charge 1 negative

   (b) 3 negative 2 positive, net charge 1 negative

   (c) 1 negative 1 positive, net charge zero

2 The primary structure of a protein molecule is which of the following?

   (a) The list of what kind of amino acids are present.

   (b) The number of acidic and basic amino acid side-chains.

   (c) The exact order in which the amino acids follow one another.

   (d) The number of each kind of amino acid that is present.

   (e) The overall shape of the molecule once it is folded.

TOPIC 5

Chemical transformations in the living organism: metabolism

Body chemistry

Looking at our body's chemistry in a blunt overall fashion, consider what we regularly take in: water, air containing 21% oxygen, various tasty foods. What we put out, on the other hand, from our various orifices (sweat, grease, wax, urine, faeces, mucus, etc. and also air containing less O_2 and more CO_2) is bewilderingly different as well as less appealing. It is obvious that a lot of complex chemical transformation is going on. What are we up to and why and how? Similarly, and more attractively, consider an infant guzzling milk. Part of this too turns into nappy contents, but the rest is turned into growing the infant – again seemingly a piece of chemical magic.

Collectively all these chemical conversion processes in the body are referred to as **metabolism**. Each such process can be written down as a chemical equation describing the net conversion. School biology, for example, is very fond of the equation:

$$C_6H_{12}O_6 + 6O_2 \rightarrow 6CO_2 + 6H_2O$$

This summarises the conversion of glucose and oxygen to carbon dioxide and water in cellular respiration. However, it is misleading in two respects. First of all, as we shall see in Topics 13–15, even as a summary it is incomplete because in a very real sense it leaves out the whole point of the process. There are other chemicals involved in the process that undergo a net transformation and ought, of course, to appear in the equation. More immediately relevant to the present point, however, it looks as if it is one gigantic reaction. In fact, at no point in our cells does a glucose

Pain-Free Biochemistry   Paul C. Engel
© 2009 John Wiley & Sons, Ltd

molecule simultaneously collide with six oxygen molecules in a molecular bonfire! The net conversion shown is the cumulative outcome of a sequence of about two dozen consecutive reactions.

Reaction sequences

Such sequences of many reactions, one after another, are known as metabolic pathways, and the chemical compounds formed step-by-step along the route are known as metabolic intermediates. Biochemists are fond of wall charts of these pathways, which add up to a truly intimidating spider's web of complexity. Most of us know only a fraction of this complexity, and for you as a health worker it certainly is not essential to know the details. On the other hand, you do need to keep an eye on the overall processes. Also we need to think about some general principles. First of all, various pathways diverge and converge and intersect, which means that individual compounds may have a choice of several possible reactions leading into different pathways. In addition, we have to bear in mind that there are also a large number of chemical reactions that a chemist might see as a theoretical possibility but do not actually occur in the living organism. What decides which reactions occur and what gives order to the whole situation? The answer, in a word, is **enzymes**, the subject of our next topic.

Self-test MCQs on Topic 5

1 What is a metabolic pathway?

(a) A carefully designed diet taking proper account of the body's needs.

(b) A way of breaking down foodstuffs to supply energy.

(c) A route for making essential cell components from simpler starting materials.

(d) Any sequence of biochemical reactions for synthesis or degradation.

2 From a biochemist's point of view, what is wrong with the following overall equation for the process of respiration?

$$C_6H_{12}O_6 + 6O_2 \rightarrow 6CO_2 + 6H_2O$$

(a) The rise in water level could drown the cell.

(b) The equation misses out most of what happens.

(c) We know that there are no gas bubbles formed.

(d) The numbers are wrong so the equation is not properly balanced.

3 From a biochemist's point of view what is wrong with the following overall equation for the process of respiration?

$$C_6H_{12}O_6 + O_2 \rightarrow CO_2 + H_2O$$

(a) The rise in water level could drown the cell.

(b) We know that there are no gas bubbles formed.

(c) The numbers are wrong so the equation is not properly balanced.

(d) Glucose does not react with oxygen.

Reactions, catalysts and enzymes

Catalysts

As mentioned earlier, the fact that a reaction is theoretically possible does not mean that it will necessarily happen at an appreciable rate. We all know, for example, that a candle can burn. This is a chemical reaction of the wax with oxygen in the air. Once it starts it will keep going, but you would be very surprised if the candle lit spontaneously. It needs some encouragement, in the form of strong heat. Other substances, such as petrol or dynamite, need rather less encouragement and we handle them with corresponding respect. Nevertheless, even these still need a spark or a sharp tap. In chemistry there are, however, agents that can speed up reactions. These are called **catalysts**, and by their presence or absence they can determine which out of all the possibilities actually happens. They can make a reaction that would otherwise take many years to run its course occur in seconds. A very important feature of catalysts is that they are still there, unchanged, after the reaction has occurred. This means, of course, that the molecules of the catalyst can work over and over again, exercising their remarkable influence on thousands of reacting molecules. The other important thing to note is that, unlike heat, which speeds up reactions in an indiscriminate way, a particular catalyst will speed up only one kind of reaction, not all reactions.

Biological catalysts

Living cells make catalysts too, very powerful ones. These are known as 'enzymes' and enzymes, like the antibodies we considered in Topic 3, are proteins.

Pain-Free Biochemistry   Paul C. Engel
© 2009 John Wiley & Sons, Ltd

If a metabolic pathway involves, say, 12 consecutive reactions, normally this will mean that there is a set of 12 different enzymes, i.e. 12 different specialised protein molecules, to 'catalyse' the 12 reactions. Each one is totally committed to its one reaction. This also means that the whole pathway can be blocked if any one of the enzymes is not there or somehow is switched off.

Enzymes have to be exceedingly potent catalysts because the living cell is deprived of several of the tricks available to the chemist in the laboratory or factory. Chemists can speed up their reactions by using high temperatures, high pressures and extremes of pH. Human cells work at 37°C, at normal atmospheric pressure, and close to pH 7, and nevertheless their enzymes are able to speed up reactions under these conditions. (Often, if one tries to use an enzyme at higher temperatures or extremes of pH, it will not work because the enzyme protein molecule is fragile and very often is irreversibly altered and inactivated – think of what happens to egg white, also mainly protein, when it is cooked.) Exactly how much they speed things up depends on the individual reaction and enzyme, and in general simple reactions are speeded up more than complicated ones, but the effect can be anything from a millionfold to a million millionfold (10^{12}-fold). Perhaps the most remarkable statistic is that a single molecule of one of the star-performing enzymes may process as many as a million molecules of its target chemical substance in a single second – a million individual reactions!

Even without getting to grips with specific metabolic pathways, we can start to see how a cell might keep some order in the chemical jungle. It can select which pathways will or will not go simply by which enzymes it makes available. Particular enzymes might or might not be there and, even if they are there, perhaps their activity could be subject to something like a 'volume control'. As it turns out, living cells use both these stratagems for regulating metabolism as we shall see in later Topics.

Self-test MCQs on Topic 6

1 Which of the following best describes the role of enzymes in metabolism?

(a) Each cell has its own enzyme molecule, which controls the metabolism in that cell.

(b) Each cell has many different kinds of enzyme molecule to speed up the many different reactions.

(c) We find enzymes in active tissues but in more sluggish types of cell ordinary chemistry takes its course.

(d) The main role of cellular metabolism is to produce enzymes.

2 What effect will an enzyme have on a reaction?

  (a) It will make it go faster.

  (b) It will make it go further.

  (c) It will make it go for longer.

  (d) All of the above.

3 Which of the following statements are true?

  (i) The drawback to enzymes is that they only work to their full potential at high temperatures.

  (ii) Although enzymes are highly specific, unfortunately they do not offer the efficient, potent catalysis achieved by typical chemists' catalysts.

  (iii) Although enzymes are extremely potent catalysts, they are somewhat fragile, unstable molecules.

  (iv) An enzyme offers a cell the great advantage of being able to catalyse a wide range of different reactions.

  (a) (iii) and (iv)

  (b) (iii) only

  (c) (i) and (ii)

  (d) (ii) only

  (e) None of the statements is true.

TOPIC 7

Specificity, saturation and active sites

Selectivity

In order to do their job efficiently, like traffic police, enzymes need to be extremely selective and recognise the right chemicals and do the right things to them. Once again, this task of molecular recognition is a job for a protein, and enzymes are indeed proteins.

Rates of enzyme-catalysed reactions

How do enzymes work? We can sneak up on the answer to this question by taking a closer look at how they behave. What they are there for is to increase rates of chemical reaction. How do we define a rate of reaction? If we imagine a reaction in which substance A is converted into substance B without any help from enzymes, then we could measure the concentration of substance A (in biology this will usually mean the concentration in solution) at timed intervals, and we would expect this concentration to decrease gradually, possibly slowly, possibly rapidly, depending on the actual substance and on the conditions – temperature, etc. If we are also able separately to measure the concentration of B, it should gradually increase, and, if the chemistry is such that one molecule of A is converted into one molecule of B, then we should expect the increase in concentration of B to exactly match the decrease in concentration of A (Fig. 7.1). The time course of these two opposite processes is what gives us the rate of reaction. The slope of the graph at any time point tells us how fast the reaction is going, and typically we would measure the rate at the start, which is the fastest rate, since the reaction gradually slows down as it progresses.

Pain-Free Biochemistry   Paul C. Engel
© 2009 John Wiley & Sons, Ltd

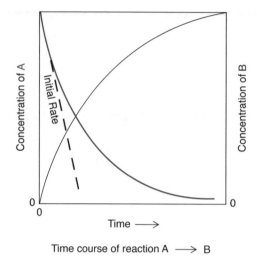

Figure 7.1 The timecourse of a reaction, showing the concentration of the starting material A decreasing as that of the product B increases.

How does reactant concentration affect the rate?

Just as you can take your coffee without sugar, with one spoonful or two, the actual starting concentration of substance A is something we can control. Suppose we decide to double that starting concentration. What happens to the rate of reaction? The rate is really a statistical result: any molecule of A will have a certain chance of undergoing reaction in any particular period of time, and, if there are twice as many molecules there, the rate should be double; if the concentration of A is ten times higher, the rate should also be ten times higher, in strict proportion (Fig. 7.2a). This is what chemists would refer to as a 'first-order dependence' of the rate on the concentration of the reacting substance. The reason for mentioning this possibly obvious-seeming idea is that this is *not* how enzyme-catalysed reactions behave. Back in the 1890s it was noticed that, if you measured the rate of a reaction catalysed by an enzyme and did so for different, gradually increasing concentrations of the reacting substance, then, at first, with low concentrations, the rate would increase in proportion, but gradually, as higher concentrations were used, there was less and less of an increase, until finally the rate approached an upper limit and would go no faster no matter how much more reactant was added (Fig. 7.2b).

The idea of a specific enzyme 'active site'

On the other hand, the measured rate *was* strictly proportional to the concentration of the enzyme. This pattern of behaviour gave us a big clue to the way enzymes

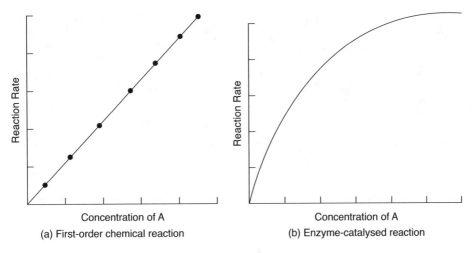

Concentration of A
(a) First-order chemical reaction

Concentration of A
(b) Enzyme-catalysed reaction

Figure 7.2   (a) 'First-order' dependence of rate on the concentration of the reacting substance, something that is frequently seen in chemistry. (b) Typical behaviour of an enzyme-catalysed reaction. At low concentrations of the reacting substance it shows approximately first-order behaviour, but as the concentration increases the curve bends over and gradually approaches 'zero-order' behaviour – no increase in rate even though the concentration increases.

work even before it was known that they are proteins. Let us imagine that each enzyme molecule has a slot on its surface, which exactly fits the target reacting substance, substance A in our argument above, in rather the way that a glove fits a hand (Fig. 7.3). If we start off with substance A in solution and no enzyme at all, all the A molecules will be experiencing the same conditions. Very likely these molecules will not react at any rate fast enough to measure. If we now add some enzyme molecules to the solution (typically a tiny number of molecules compared to the molecules of substance A), most of the A molecules will still be floating free in solution and reacting very, very slowly, if at all, but at any given time a few will have managed to find the slot on an enzyme molecule and will react very much

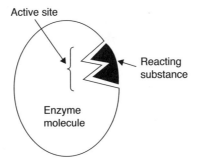

Active site

Reacting
substance

Enzyme
molecule

Figure 7.3   The idea of an enzyme 'active site', with a slot tailor-made for the right reacting substance.

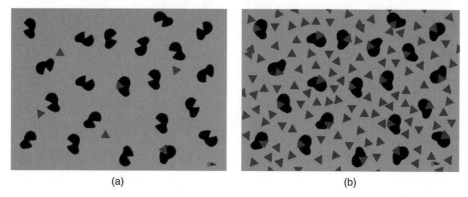

(a)   (b)

Figure 7.4 Enzyme active sites explain the behaviour shown in Fig. 7.2b. In Fig. 7.4a there are still plenty of unoccupied enzyme active sites and so the rate of reaction goes up almost in proportion to the concentration of substance A (red triangles). In Fig. 7.4b the enzyme active sites are virtually all filled and, no matter how many more molecules of A you add, the rate of reaction cannot increase.

faster. They will react so much faster, in fact, that for any appreciable addition of enzyme the rate we measure will be overwhelmingly due to the fraction of A molecules (even though it may be a very small fraction) that is temporarily attached to the enzyme molecules. So now, as we increase the concentration of A, what happens is that more and more of the enzyme slots become occupied and so the reaction will go faster and faster (Fig. 7.4a). Eventually, however, they are almost all occupied (Fig. 7.4b), and as fast as a molecule of B is formed and drops out of the slot another A molecule jumps in. (You may be doubtful about the jumping molecules! A useful concept, if so, is the idea that in an apparently still solution, if you could see the individual molecules they would be buzzing around like insects. The hotter the solution the faster they buzz!) So once all the enzyme molecules are working flat out, things can go no faster. You can think of people (molecules of A) trying to get home after a night on the town; they could walk it, but a taxi (enzyme molecule) would get them safely home so much faster. At first there are taxis waiting at the ranks, but, as more and more people come pouring out, all the taxis are busy and no matter how long the queues become people are not going to get home any faster! On the other hand, if the taxi companies suddenly put on some more drivers and released more taxis from the garage (more enzyme molecules), that of course would speed things up.

The idea of the slot on the enzyme molecule is a very powerful one and has illuminated many other processes. It accounts for the extreme selectivity of enzymes – the wrong hand will not fit the glove – and it is really the same idea as we already met in Topic 3 to account for antibody recognition. It depends on the ability of individual types of proteins to adopt very precise elaborate shapes. In the present case, the recognition molecule, the enzyme, is not content with recognition alone: it sets about rapidly altering its chemical captive. Clearly the 'slot' in an enzyme, which biochemists refer to as the enzyme's 'active site', has some special properties.

They are really beyond our scope in this book and all you need to remember is that, by very briefly holding a molecule (or sometimes more than one molecule) in a firm grasp, the enzyme is able to deploy all its chemical tools: everything is positioned exactly as it needs to be for maximum efficiency.

Self-test MCQs on Topic 7

1 What is meant by the rate of a reaction?

(a) It is the concentration of the reaction product.

(b) It is the time taken for the reaction to start.

(c) It is the time interval from the beginning of the reaction to the end.

(d) It is the change in concentration of reactants per unit time.

2 What would happen to the rate of an enzyme-catalysed reaction if you were able to double the number of enzyme molecules?

(a) The rate should double since there are twice as many sites of catalysis.

(b) The rate would halve since the reactant molecules have to be split among more enzyme molecules.

(c) The rate ought not to change since the reaction is already catalysed.

(d) It is impossible to predict without knowing what the reaction is.

3 What would happen to the rate of an enzyme-catalysed reaction if you were able to double the concentration of one of the reacting substances?

(a) The rate would decrease since the enzyme molecules have to work twice as hard.

(b) The rate should not change if it is purely an enzyme-catalysed reaction.

(c) The rate would be expected to double.

(d) It is impossible to predict without knowing how fully occupied the enzyme molecules already are.

Structure of metabolism: anabolism and catabolism

Why do we need to transform substances?

Having defined metabolism and enzymes, we can look closer and ask what metabolism is trying to achieve. First of all, living cells are made up of a large number of complex chemical substances. These all have to be made, with the exception of some of the smaller molecules that we are able to take in and use ready-made. Growth, repair and cell division all require many different new molecules. Reactions to make them are referred to as **biosynthesis** or **anabolism** (Fig. 8.1), but what from? Plants can start from scratch with CO_2 and water, and simple mineral salts, but we cannot. Therefore, we need more complex building blocks. Where from? Building blocks can come either directly from food or else in some cases from food-derived stores. These reactions to generate building blocks make up **catabolism**. For example, we eat meat that contains proteins very similar to our own muscle proteins, but we cannot simply use the intact proteins. We break them down into amino acids (Topic 23) and then can reassemble these to make our own proteins (Topic 35). (Indeed if we did not, the foreign proteins would certainly cause an immune reaction, as discussed in Topic 3.) This perfectly normal anabolic process has acquired a good deal of notoriety through the exploits of athletes, who try to pile on unnatural amounts of muscle by using *anabolic* steroids.

However, the relationship of catabolism and anabolism is not merely that of one process providing the raw materials for the other. They are linked in another fundamental way, in that catabolism also provides the driving force that makes biosynthesis possible. We shall explore this process, usually described as 'energy metabolism', in the sections immediately following this one.

Pain-Free Biochemistry   Paul C. Engel
© 2009 John Wiley & Sons, Ltd

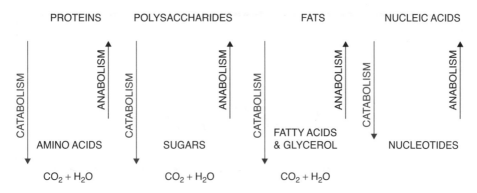

Figure 8.1   Catabolism and anabolism. The diagram indicates the interconversion of some of the main categories of biological molecule and their various building blocks.

Self-test MCQ on Topic 8

Which one of the following is the correct statement?

(a) Anabolism is made up of metabolism (breakdown) and catabolism (synthesis).

(b) Metabolism is made up of catabolism (synthesis) and anabolism (breakdown).

(c) Anabolism is made up of catabolism (breakdown) and metabolism (synthesis).

(d) Metabolism is made up of anabolism (synthesis) and catabolism (breakdown).

Equilibrium

See also Appendix 4.

Where does a reaction stop?

To understand the next crucial aspect of metabolism, bioenergetics, we need to go back to chemistry and chemical reactions. We have seen that a chemical reaction might or might not occur at a measurable rate depending, for example, on whether there is an efficient catalyst or whether the chemicals involved are intrinsically reactive. However, let us suppose we have two different chemical compounds A and B, which can react together to form two more compounds, C and D, and there is an efficient enzyme to make it happen. We shall come across many examples in the 'Topic' sections to come, but a typical example would be the 'transaminase' reaction below, which we shall meet 'for real' in Topic 23.

glutamic acid + oxaloacetic acid ↔ α-ketoglutaric acid + aspartic acid

We do not need to concern ourselves for the moment with the detailed chemical structure that goes with each of these names. This is a very symmetrical sort of reaction in which an amino acid, glutamic acid (A) gives up its amino ($-NH_2$) group to a receiving α-ketoacid, oxaloacetic acid (B). This converts glutamic acid into an α-ketoacid, α-ketoglutaric acid (C), and the oxaloacetic acid, sporting its new amino group, has now become the amino acid aspartic acid (D). If you do not follow that, do not worry; stick with A, B, C and D because what we need to think about here is entirely general.

We have defined our reaction by an equation that tells us that one molecule of A reacts with one molecule of B to form one molecule each of C and D, but there are still other questions to ask about this reaction. If we mix equal concentrations of A and B:

1 Can we state how far the reaction will go?

2 Will it go until all the A and B are used up?

3 Suppose instead we mixed C and D, would they react?

The answers are:

1 How far the reaction will go depends on what A and B are, but in principle for any particular pair we can work it out exactly.

2 No. Some reactions will go very close to completion, but there will always be a bit left, whether it is 50% or 0.001%.

3 Yes. All chemical reactions are in principle reversible. It might or might not go very far, but it would certainly go some distance, making A and B in the process.

Equilibrium

As A and B get used up, the forward reaction inevitably goes more slowly; as C and D accumulate, the reverse reaction starts up and goes gradually faster and faster. Just measuring the changing concentrations will make it appear that the reaction is moving steadily in one direction. In fact, it will be going in both directions simultaneously, and what we measure is the net result. Eventually, as one slows down and the other goes faster, they fight one another to an apparent standstill. How far the reaction goes, therefore, is determined by the point at which the reverse reaction exactly balances the forward reaction. This point will vary from reaction to reaction and is governed by the '**equilibrium constant**' (Appendix 4). Going back to question 3 above, therefore, starting with, say, 1 mM C and 1 mM D will give exactly the same final equilibrium composition of A, B, C and D as starting with 1 mM A and 1 mM B (Fig. IV.1). It is important to realise that the *equilibrium constant is a property of the chemical compounds* and cannot be altered by the presence or absence of an enzyme. An enzyme merely influences how rapidly the reaction approaches equilibrium.

Driving conversions in the desired direction

Let us suppose that a particular reaction goes only 10% of the way to completion but that we would really like all the B to be used up. Are we stuck with 10%? The

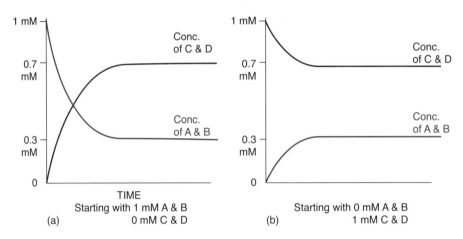

Figure IV.1   Approaching reaction equilibrium. If A + B ↔ C + D , then you should end up with exactly the same mixture whether you start with 1 mM A and 1 mM B or with 1 mM C and 1 mM D.

answer is that we can either push the reaction by piling in high concentrations of A or pull it by removing C or D, or both, as fast as they are formed (so that the reverse reaction cannot get established). There is also an alternative that, as we shall see, living things use extensively. We might be able to carry out a different 'priming' reaction on A, converting it into a compound that will still react with B and convert it into D but now with a much more favourable equilibrium constant.

Self-test MCQs on Chemistry IV

1 Suppose we have an enzyme that catalyses a reaction in which compound A reacts with compound B. Compound A is converted into C and compound B is converted into D:

$$A + B \leftrightarrow C + D$$

We start off with 10 mM A and 2 mM B mixed in solution. We then add the enzyme and measure the concentration of D, as it increases from zero. Eventually, the concentration stops changing and D is measured at 1 mM. What will be the concentrations of A, B and C?

(a) 10 mM, 1 mM, 0 mM

(b) 0 mM, 2 mM, 10 mM

(c) 9 mM, 1 mM, 1 mM

(d) 1 mM, 1 mM, 9 mM

2 Suppose we are not happy with the reaction above because we were hoping to get B more completely converted to D. How could we treat the reaction mixture to achieve this? (There could be more than one answer to this question.)

(a) Add more enzyme.

(b) Be patient and wait for the reaction to start up again.

(c) Add a better enzyme.

(d) Add a lot more of compound A.

(e) Add a lot more of compound C.

(f) Remove compound C as fast as it is formed.

(g) Remove compound D as fast as it is formed.

(h) Boil the mixture.

TOPIC 9

Catabolism: degradation vs energy metabolism

Energy metabolism – ATP

As we have seen, some catabolic reactions simply generate building blocks: molecular Lego. A considerable part of food breakdown, however, is devoted to a dual function in which provision of building blocks is combined with an energy requirement A large proportion of the fat and **carbohydrate** that we metabolise is converted to CO_2 and water. We cannot consider these as 'building blocks' (although a plant would). We do it in order to produce two commodities – **ATP** and 'reducing power'.

ATP (Fig. 9.1), adenosine *tri*phosphate, is often spoken of as the 'universal energy currency of the cell'. This 'currency' is spent by chemically converting it to ADP, adenosine *di*phosphate (two phosphates) and phosphate (or sometimes, more extravagantly, to AMP, adenosine *mono*phosphate and pyrophosphate). In reverse, it is earned and saved by converting ADP and phosphate back to ATP (see Appendix 5 for more about phosphate). The reason we refer to it as a *universal* currency is that different energy foodstuffs are *all* broken down in such a way that they drive the formation of ATP; likewise dozens of processes in our bodies that need an energy push are so constructed that phosphate is split off ATP as they occur. ATP is thus a way of storing chemical energy so that it can be deployed wherever it is needed to drive processes that would otherwise not occur because of unfavourable equilibrium constants. You can roll a car down a hill but cannot expect it to roll back up again. On the other hand, if you turn on the engine and use up some petrol, it becomes easy to drive the car back up the hill. In a similar way, with a little help from enzymes, proteins can be broken down into the separate amino acids (necklace into beads). This is the downhill process, which goes very readily

Pain-Free Biochemistry   Paul C. Engel
© 2009 John Wiley & Sons, Ltd

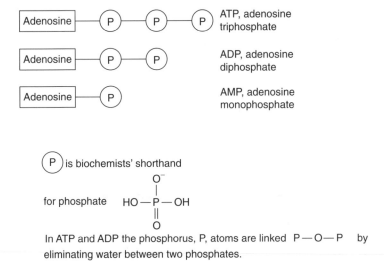

Figure 9.1   ATP (adenosine triphosphate).

because the equilibrium constant lies very far over towards breakdown. For exactly that same reason we cannot make proteins simply by reversing that set of reactions, i.e. by mixing the amino acids, adding the enzyme and hoping for the best. (From what we said in Chemistry IV, we *could* in theory reverse it but only by either piling in enormous extra concentrations of the amino acids or removing the product protein as fast as it was made in order to keep its concentration at vanishingly small concentrations. Neither of these is a practical solution to the problem for our cells.) That would be an uphill process, and, like the car rolling up the hill, impossible without help. Instead, the cell uses a different set of reactions that incorporate splitting of ATP molecules, equivalent to using some petrol in the car.

In the opposite direction, there are lots of breakdown (catabolic) reactions that could easily run out of control, with very favourable equilibrium constants, but metabolism is cunningly constructed to divert some of these processes down different channels so as to make ATP in the process. To use an analogy from everyday life, we have national electricity grids that nowadays might have inputs from coal or gas-fired power stations, hydroelectric turbines, wind power, solar panels, nuclear power stations, and many other physical sources of energy but all separately engineered to produce electricity. We also have a multitude of processes that rely on that electricity – cooking the dinner, lighting our streets, running our factories, ironing the clothes, etc. Because power supply and power demand keep fluctuating, there has to be a way of storing excess electricity. In real life, power engineers have a number of different ways of solving that problem, but for the sake of our analogy, since it is a chemical solution, let us focus on giant lead accumulators, which are huge batteries. The chemical conversion in the batteries would be equivalent to our interconversion of ATP and ADP: a range of different inputs can drive the process in one direction, and then, by running back the other way, the batteries, like ATP, can drive a whole diverse set of output processes.

At the beginning of this section, we mentioned that, as well as ATP, catabolic reactions tend to generate 'reducing power'. This is another important linkage between catabolism and anabolism, helping one process to supply and drive the other (see Topics 15 and 18), but in order to understand this, we need to take another short detour to think about the basic chemistry of **oxidation** and **reduction**.

Self-test MCQs on Topic 9

1 ATP is important in metabolism for which one of the following reasons?

(a) Being very chemically stable, it is an ideal compound for long-term energy storage.

(b) It can be used to release large amounts of heat and this heat drives biosynthesis.

(c) Of all the energy-giving foodstuffs it is most easily absorbed in the gut.

(d) Being made by catabolism and used by anabolism, it allows one process to drive the other.

2 Which of the following is true?

(a) A catabolic reaction by definition forms ATP from ADP.

(b) Some catabolic reactions form ATP from ADP.

(c) Catabolic reactions yield building blocks and are distinct from the ATP-forming reactions.

(d) Catabolic processes are driven forward by net conversion of ATP to ADP.

CHEMISTRY V

Oxidation and reduction

Combustion and combination with oxygen

We have already mentioned the metabolic conversion of glucose to CO_2 and water: the equation for respiration with this foodstuff involves adding O_2 molecules, as indeed it would if we were using any other energy-supplying food material. So it is time to consider the significance of oxidation. Long before biochemistry could be properly studied, chemists in the eighteenth century were giving their attention to what exactly goes on when substances burn in air. This led to the realisation that air is not just a single pure substance but rather a mixture of several gases, with oxygen making up about one fifth of all the molecules. Combustion usually involves a substance reacting with this oxygen. This can happen in one of two ways. If you heat copper powder to a high temperature in air, it goes black and it also becomes *heavier* because it has reacted with oxygen from the air to form copper oxide, CuO.

$$2Cu + O_2 \rightarrow 2CuO$$

(Note that we have to double up on the copper atoms since the oxygen molecule contains two atoms.)

Oxidation by removal of hydrogen

This process of adding oxygen is called **oxidation**. However, if we think of burning a hydrocarbon substance (gas in a gas cooker, petrol in a car engine and a candle) it can burn in two ways. If there is a good supply of oxygen we see complete combustion, and all the carbon is converted to CO_2 and all the hydrogen to H_2O. If there is not enough oxygen we tend to see smoke. In this case the hydrogen is still converted into water, but some of the carbon remains as black carbon. We may

Pain-Free Biochemistry   Paul C. Engel
© 2009 John Wiley & Sons, Ltd

think of the oxygen pulling hydrogen atoms off the carbon skeleton. *This is still seen as oxidation even though in this case the carbon atoms do not end up linked to oxygen.* This is shown below in two possible equations for combustion of propane (camping gas). There would also be equations for situations intermediate between these two extremes.

$$C_3H_8 + 5O_2 \rightarrow 3CO_2 + 4H_2O \text{ (complete combustion)}$$
$$C_3H_8 + 2O_2 \rightarrow 3C + 4H_2O \text{ (partial combustion)}$$

Chemistry has extended the definition of oxidation further. Removal of hydrogen atoms by oxygen is seen as oxidation, but, going further now, removal of hydrogen atoms by other substances is also regarded as oxidation. We shall see in due course that in biochemistry, pairs of hydrogen atoms are passed between different oxidising substances rather like the baton in a relay race, and are indeed finally passed on to oxygen at the end of the line – the 'terminal acceptor'. Thus, *either* addition of oxygen atoms *or* removal of hydrogen atoms counts as oxidation.

Removal of oxygen

Going back now to our copper oxide, we can consider the reverse process, which is called **reduction**. If you were to heat the copper oxide again and pass hydrogen over it, the black copper oxide would gradually convert back to metallic copper.

$$CuO + H_2 \rightarrow Cu + H_2O$$

Just as oxygen can remove hydrogen atoms as water, so hydrogen can remove oxygen atoms as water.

Hydrogenation

To complete the picture, just as oxidation might or might not involve a molecule actually taking up the oxygen, so also hydrogen, which is not incorporated into the CuO molecule in the reaction above, can sometimes be incorporated in other reduction reactions. For example, in the food industry, **unsaturated** fats (see Chemistry IX) are converted into saturated fats by catalytic hydrogenation – pairs of hydrogen atoms are incorporated, and this is regarded as a chemical reduction. As mentioned above, many biochemical oxidations involve removal of pairs of hydrogens; conversely many biochemical reductions involve donation and incorporation of pairs of hydrogens.

Oxidation involves reduction

We should note that any reduction reaction is necessarily also an oxidation! We showed the copper oxide reaction above to emphasise that the CuO was reduced by the hydrogen, but by the same token the hydrogen is oxidised by the CuO. Each donor is always matched by an acceptor. It takes two to tango.

Change of valency

Finally let us stretch our definition a little further. A number of the heavier metal atoms, for reasons of atomic structure that we need not worry about, have a choice of valency states. Copper and iron are two of these. Copper (Cu) may have a valency of either 2 or 1. Iron (Fe) may have a valency of either 2 or 3. We have already seen Cu, with a valency of 2, combined with O, also with a valency of 2, to make a one-to-one compound CuO. To be explicit about the valency we should call this compound cup*ric* oxide. However, in a valency state of 1, it requires *two* Cu atoms to satisfy oxygen's valency of 2, and so we have Cu_2O, which is distinguished from CuO by the name cup*rous* oxide. This is a different compound made up of the same two kinds of atoms but in different proportions. It is possible to convert one into the other by an oxidation reaction:

$$2Cu_2O + O_2 \rightarrow 4CuO$$

Receiving and donating electrons

Now, if we extend this thinking, we can consider the interconversion not just of copper oxides but of ionic copper compounds, like copper chlorides (cupric chloride, $CuCl_2$, and cuprous chloride, $CuCl$). Suppose we have a mixed solution of cuprous chloride and hydrochloric acid (HCl). Both compounds will split into their ions, Cu^+, Cl^-, H^+ and another Cl^-. If the Cu^+ were oxidised to the cupric state, Cu^{2+}, it could 'claim' both the chloride ions. For this to happen, the Cu^+ must lose a negative charge, an electron. Who better to lose it to than the hydrogen ion, H^+! If a pair of cuprous ions are oxidised to cupric ions by a pair of H^+ ions, the Hs can pair up to form hydrogen gas H_2.

What we have sneaked up on here is another definition of changes to be included under the oxidation–reduction umbrella. Removal of an electron can also be considered an oxidation, and donation of an electron is a reduction. Although this might seem a bit esoteric, the ability to alternate between reduction by electrons and reduction by hydrogen atoms turns out to be very important in bioenergetics (see Topic 15).

So to summarise, oxidation can involve addition of oxygen atoms, removal of hydrogen atoms or removal of electrons; reduction, conversely, can involve removal of oxygen atoms, addition of hydrogen atoms or addition of electrons.

Self-test MCQs on Chemistry V

1 In the equation $CH_4 + 3O_2 \rightarrow CO_2 + 2H_2O$, which of the following is true?

  (i) Oxygen is being reduced by methane.

  (ii) Oxygen is being oxidised by methane.

  (iii) Methane is being reduced to carbon dioxide.

  (iv) Methane is being oxidised to water and carbon dioxide.

  (a) Both (i) and (iv)

  (b) Only (i)

  (c) Both (ii) and (iii)

  (d) Only (iii)

  (e) Both (i) and (iii)

2 Oxidation includes which of the following?

  (a) Gaining electrons, gaining oxygen atoms or losing hydrogen atoms

  (b) Losing electrons, losing oxygen atoms or gaining hydrogen atoms

  (c) Losing electrons, gaining oxygen atoms or losing hydrogen atoms

  (d) Gaining electrons, losing oxygen atoms or gaining hydrogen atoms

TOPIC 10

Oxidation and reduction in metabolism

Biological oxidation and oxidative cofactors

Our catabolic processes broadly involve oxidation. Conversely, our biosynthetic processes involve reduction. In theory, all the hydrogen atoms that get stripped off sugars and fats as we break them down could be handed over straight away to oxygen, the 'terminal oxidant', making water in a reaction similar to the combustion of propane we considered in Chemistry V. If so, a lot of energy would be released as heat (as indeed it is when we burn propane!). Biologically this would mean that the energy was wasted (unless needed to keep us warm!) because, once the energy has been let loose into the surroundings, there is no way of ever gathering it up again to drive our energy-requiring processes. What living things are trying to do is to trap as much of the energy as possible in a useable, controllable form, to drive biological processes. So, instead, the cell has a set of intermediate oxidising substances, in particular NAD^+, $NADP^+$ and FAD (the letters in these names are *not* symbols for atoms; they are abbreviations for the long names of larger chemical structures, which are rather a mouthful if you have to say them often! See Box 3). These substances intercept the hydrogen atoms and get reduced in the process (NADH, NADPH and $FADH_2$). These reduced **cofactors** give the cell two choices. Either the hydrogens can be passed on from one cofactor to another in a series of reduction reactions that are used to drive the formation of ATP (just as a stream can be diverted to drive a set of water wheels; see Topic 15) or hydrogens from the breakdown of one foodstuff can be used to drive the reduction reactions in the biosynthesis of another substance. For example, NADPH from oxidation of glucose (see Topic 28) may drive the reduction reactions to make fat (see Topic 27).

Pain-Free Biochemistry   Paul C. Engel
© 2009 John Wiley & Sons, Ltd

Turnover of cofactors

These cofactors are all made from substances that our cells are unable to make for themselves. This means we have to take them in through our diet. These are vitamins (Box 3). The cofactors are correspondingly present in our cells at relatively low concentrations, and yet they can process a flow-through of large amounts of oxidisable foodstuffs. How is this possible, since we know that one molecule oxidised means another molecule reduced? The answer is that the cofactors work a little like cog wheels in a machine, turning over and over rapidly. As fast as molecules of,

---

Box 3   Vitamins

The first major vitamin to be recognised in terms of its influence was vitamin C. This emerged as it was realised that scurvy, a disease of sailors in the days of journeys of many months in sailing ships, was the result of a severely restricted diet, and that it could be relieved or prevented with plant-derived material. Jacques Cartier gave his men a decoction of pine needles; James Cook took the simpler option of taking citrus fruit on board and to this day Americans refer to British servicemen as 'limeys'. It was another 150 years before the actual substance responsible for these remarkable effects was isolated from another rich source, green peppers, and identified. Indeed the word 'vitamin' was only introduced in 1912.

Most of the vitamins were gradually discovered during the first half of the twentieth century, often through the study of tropical diseases of malnutrition. Thus, beri-beri was associated with a deficiency of thiamine (vitamin B1), pellagra with a deficiency of nicotinamide, etc. Nowadays, the vitamins are generally divided into the fat-soluble vitamins, A, D and E, and the water-soluble vitamins, which include vitamins C and K and a set of unrelated compounds under the general bracket of 'B vitamins'. Vitamin K is important in making the enzymes involved in blood clotting (Topic 54). Vitamin C is crucial for making collagen, the major protein in connective tissue and bone (see also Appendix 3). The various B vitamins are required in order to supply a range of enzyme cofactors. Thus, riboflavin (B_2) is needed to make the flavin cofactors FAD and FMN, nicotinamide to make NAD^+ and $NADP^+$, pantothenic acid to make **coenzyme A (CoA)** and acyl carrier protein, thiamine to make thiamine pyrophosphate, etc. The reason we have to take in these substances, albeit in tiny amounts, in order to remain healthy is that our body chemistry cannot produce them. Many simpler organisms can make these compounds. For us it is wasteful to make enzymes for synthetic pathways that we do not require. Let someone else do the work! This policy works so long as we eat a varied and adequate diet.

say, NAD$^+$ get reduced to NADH, they pass on the reducing equivalents to other molecules so that they are straight back again as NAD$^+$ to carry out more oxidation if required.

Nicotinamide reduction

Finally, a word of clarification about the mysterious plus sign over the NAD, which disappears when the NAD gets reduced! As we can see above, when FAD is reduced, it picks up two hydrogen atoms and becomes FADH$_2$. In fact, the reduction of NAD$^+$ also involves two hydrogen atoms, but only one of them actually goes into the cofactor molecule: the N of NAD$^+$ is nicotinamide (Box 3), and it is this bit of the NAD$^+$ molecule that undergoes the chemistry (Fig. 10.1). The nitrogen atom in the nicotinamide ring carries a positive charge and when a molecule tries to donate two hydrogen atoms to the nicotinamide, they split into H$^-$ (a hydride ion) and H$^+$ (a proton). The cofactor accepts the H$^-$ and so it becomes NADH, with its positive charge neutralised by the incoming negative charge. The proton, H$^+$, is left over in free solution.

$$NAD^+ + 2H \leftrightarrow NADH + H^+$$

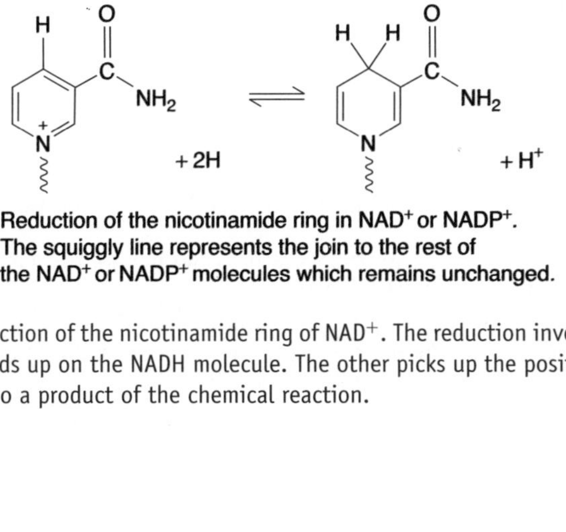

Reduction of the nicotinamide ring in NAD$^+$ or NADP$^+$.
The squiggly line represents the join to the rest of
the NAD$^+$ or NADP$^+$ molecules which remains unchanged.

Figure 10.1   Reduction of the nicotinamide ring of NAD$^+$. The reduction involves *two* hydrogen atoms. Only one ends up on the NADH molecule. The other picks up the positive charge so that a proton, H$^+$, is also a product of the chemical reaction.

Self-test MCQs on Topic 10

1 Which of the following best summarises the role of NAD^+ in energy metabolism?

  (i) Rather like gasoline, NAD^+ has a high hydrocarbon content and so its combustion yields large amounts of useable energy for the cell.

  (ii) NAD^+ provides the critical building blocks for making the ATP molecule.

  (iii) NAD^+ can be used to oxidise metabolic substrates.

  (iv) Reoxidation of NAD^+ can drive the formation of ATP.

  (v) NAD^+ is a vitamin-derived cofactor.

  (vi) NAD^+ from our diet is converted to the oxidative cofactor FAD.

  (a) (i)

  (b) (i) and (v)

  (c) (v) and (vi)

  (d) (ii)

  (e) (ii) and (v)

  (f) (iii)

  (g) (iii) and (iv)

  (h) (iii) and (v)

  (i) (iii), (iv) and (v)

2 Vitamins are only needed in trace amounts in our diet because

  (a) the cofactor molecules derived from them can be used over and over again

  (b) the efficiency of our energy metabolism means that surprisingly small amounts of 'fuel' are needed

  (c) most of the time the enzymes involved can manage without cofactor assistance;

  (d) once each cofactor molecule has been converted, e.g. reduced, it can continue in this enzymatically active state more or less indefinitely.

SECTION 2

Catabolism

Aldehydes, ketones and sugars

Carbohydrates

Our next topic is carbohydrates, one of our most important foodstuffs and also one of our two ways of storing food within our own bodies. First, we need to consider what carbohydrates are. They are **sugars** and sugar polymers (chains of sugars joined to one another). But what is a sugar?

To understand sugars we need to introduce two types of chemical 'group' that are very common in organic chemistry. (A group is an arrangement of atoms that crops up again and again.) These are the **hydroxyl** group and the **carbonyl** group.

Hydroxyl group

A hydroxyl group, as the name suggests, has hydrogen linked to oxygen, $-OH$. This leaves the oxygen atom, O, with one hand free (one spare valency), which it can use to link to (e.g.) a carbon atom. The simplest organic hydroxyl compounds, collectively called **alcohols**, are CH_3OH, methanol (very poisonous to us), and ethanol, the alcohol of alcoholic drinks (also poisonous, but less immediately, unless in very high doses) (Fig. VI.1).

Carbonyl group

A carbonyl group is a carbon atom linked to oxygen through both the oxygen valencies – a 'double bond' $)C{=}O$. This leaves two carbon valencies free, since it has

---

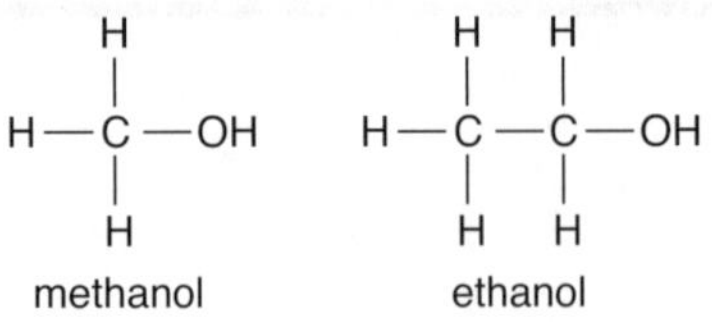

Figure VI.1   Alcohols.

four. In the simplest carbonyl compound, formaldehyde, both these valencies are occupied by hydrogen atoms (Fig. VI.2). On the other hand, the carbon valencies could be occupied by other linkages, e.g. to other carbon atoms, as in acetaldehyde (one CH_3 and one H) or acetone (two CH_3 groups), familiar to everyone as nail varnish remover, and also, on the breath, a clinical sign of uncontrolled diabetes (See Topics 21 and 44). So long as one of the two spare positions on the carbonyl C carries a hydrogen atom the compound is classed as an **aldehyde** (e.g. form*aldehyde*). If *both* positions are linked to other C atoms, the compound is a **ketone**, e.g. acet*one*.

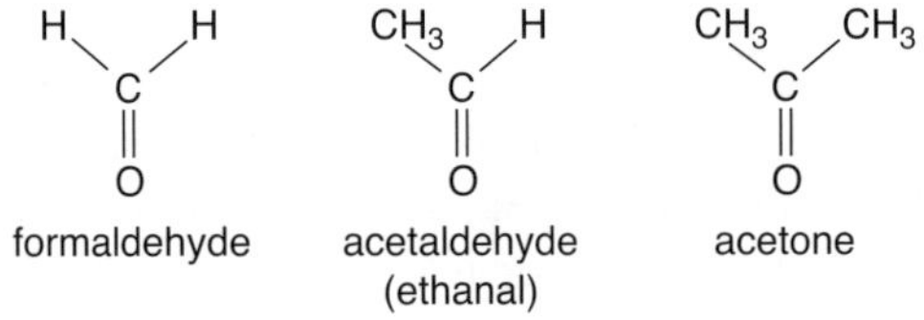

Figure VI.2   Aldehydes and ketones.

Sugars

How does this relate to sugars? Sugars are defined as carbon compounds with at least two carbon atoms linked to an —OH group and at least one carbonyl group (not linked to an —OH). This means at least three carbon atoms, and the simplest sugars are 3-carbon sugars, or **trioses**. (Chemists and biochemists use the ending -ose to name sugars – like glucose, sucrose, maltose, etc.) A triose could be an aldehyde (H—C=O on an end carbon atom) or a ketone (C=O on the middle carbon). The 3C ketone sugar is 'dihydroxyacetone' (the acetone structure above with two —OHs added on). The aldehyde sugar is 'glyceraldehyde' (Fig. VI.3).

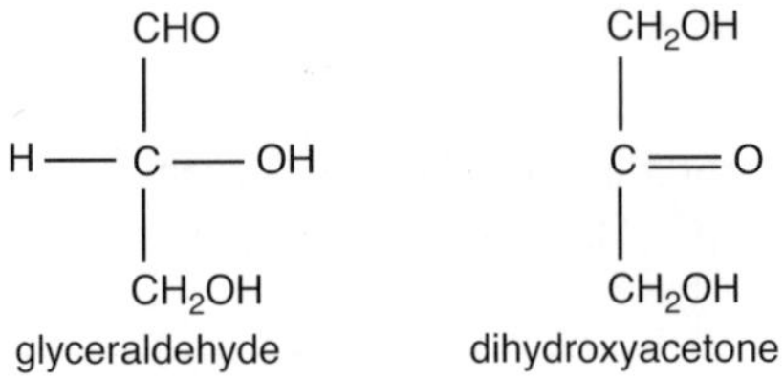

Figure VI.3   Triose sugars.

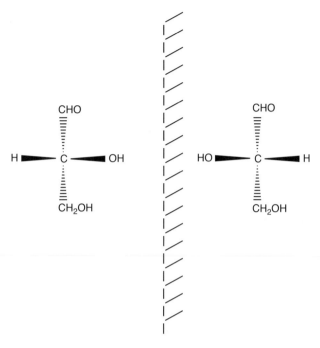

Figure VI.4   Mirror-image molecules: D- and L-glyceraldehyde.

Asymmetric carbon

There is one more complication relating to glyceraldehyde: its central carbon atom has four different partners —H, —OH, —CHO and —CH$_2$OH. The four valencies are symmetrically arranged *in 3-D space*, not on a flat surface like a page. They point to the four corners of a tetrahedron. If we arrange four different atoms or groups attached to the four positions, there are two different ways to do it, mirror images that cannot be overlapped (Fig. VI.4). If you are not convinced, try it with cocktail sticks and two potatoes! This means there are two different glyceraldehydes, D-glyceraldehyde and L-glyceraldehyde, 'right-hand' and 'left-hand' versions, respectively. Chemically they will behave identically, but remember now that biochemical systems use big molecules with complex 3-D shapes, like enzymes. These are themselves asymmetrical. As we mentioned earlier, a right-hand glove will not fit a left hand, and so also an enzyme that works for a D-compound will not accept the L-compound because it simply will not fit.

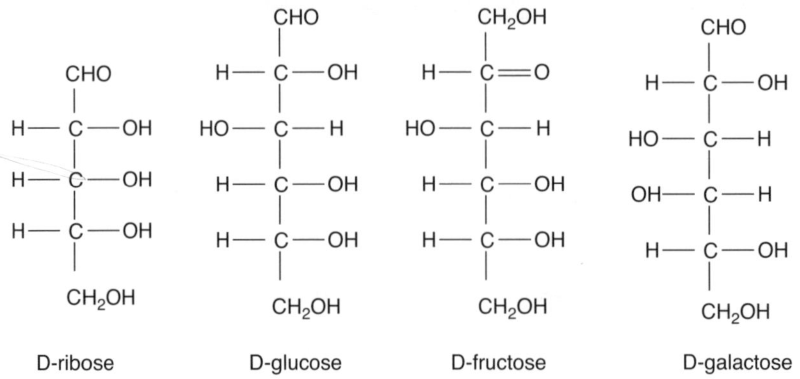

D-ribose   D-glucose   D-fructose   D-galactose

Figure VI.5   Biologically important 5- and 6-carbon sugars.

Longer sugars

One can make longer sugar molecules by putting in more carbon atoms, each carrying an H and an OH. Each time, however, this adds an extra asymmetric C atom, offering two possible choices. So while there are only two C3 aldehyde sugars, there are four C4s, eight C5s (**pentoses**), 16 C6s (**hexoses**) and so on. In fact, in spite of this huge number of possible sugars, human biology mostly restricts itself to a relatively small set of important ones. Most important are the five-carbon sugars (e.g. ribose) and six-carbon sugars (glucose, fructose and galactose) (Fig. VI.5).

Cyclisation

Molecules can twist/rotate round their bonds. This means that five- and six-carbon sugars need not be stretched out as drawn in Fig. VI.5, and in their gyrations they can curl round end to end. This makes possible a surprising feature of sugars: in chemistry —OH compounds can easily react with C=O compounds, but sugars have both OH and C=O groups in one and the same molecule, and so the C=O at one end of a sugar molecule can react with an OH at the other end – forming a ring (Fig. VI.6)! You might reasonably think that this would make a new different chemical compound but this conversion is rapidly reversible, so that the sugar flicks in and out of the ring form and both straight-chain and ring forms are regarded as features of the same sugar. In rather the same way, a convertible car might have its hood up or down, but you would still consider it the same car.

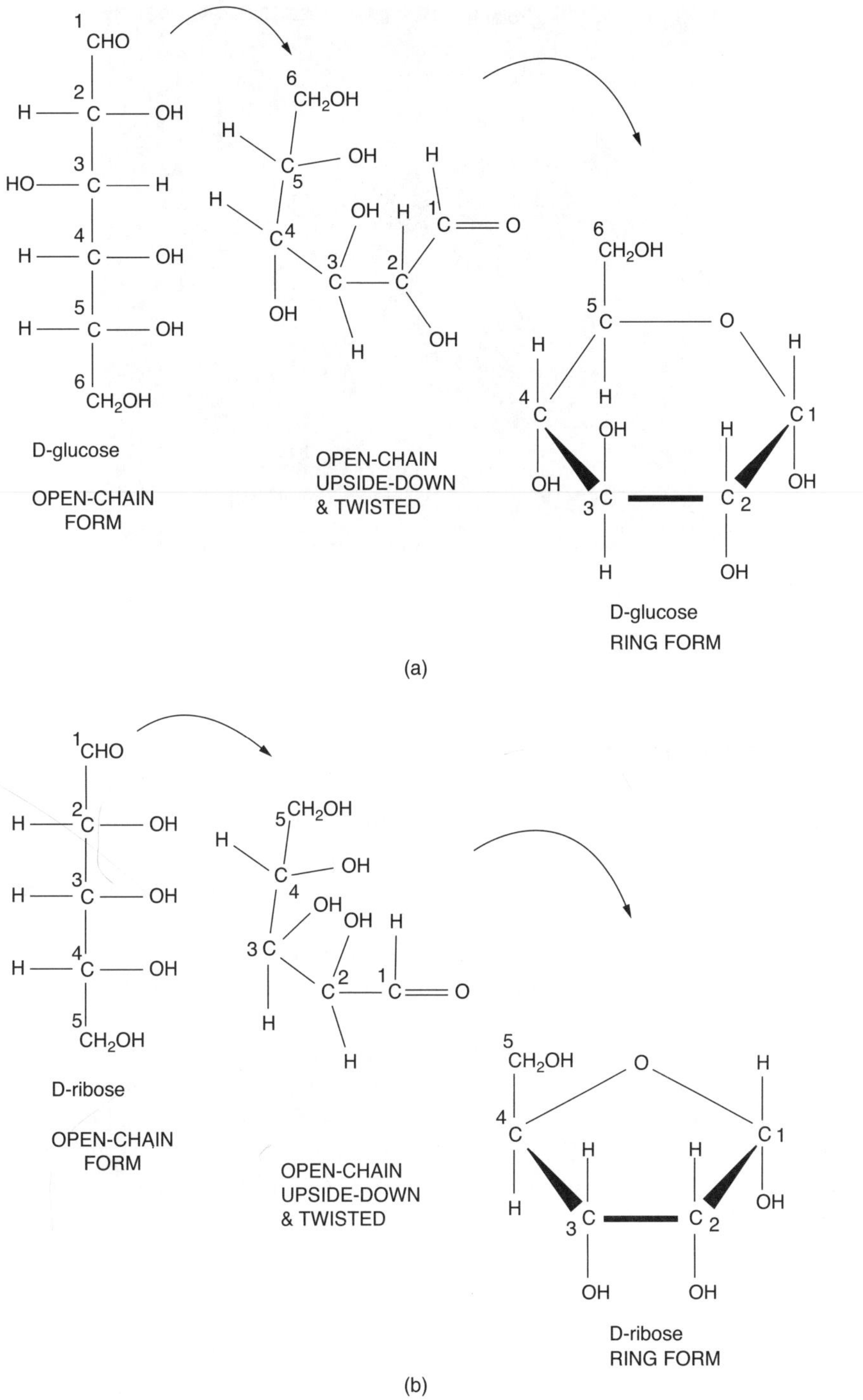

Figure VI.6   Sugar ring formation. This is a reversible process, constantly going to and fro in free solution.

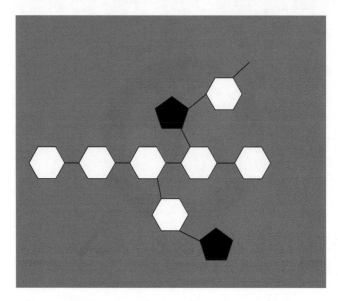

Figure VI.7   Pentoses and hexoses have several positions through which they can link to other sugars.

Joining sugar units

Carbohydrates are also classified as 'saccharides' (from the Greek word for sugar). If there is only one sugar unit then we have a monosaccharide, e.g. glucose, but sugars can be joined in chains to form disaccharides (two), trisaccharides (three) and **polysaccharides** (many). They are joined (on paper) by eliminating water (H_2O) between two —OH groups to form a **glycosidic** link. (Note: this is *not* how these reactions are carried out in the cell; it simply explains the structure.)

Unlike amino acids that have to join -amino to -carboxyl, a five- or six-carbon sugar has several —OH groups and therefore potentially several choices of position to link to another sugar, *and*, even when joined, say, between C1 of one sugar and C6 of another, both sugars still have several positions left free, so that it is possible to have branched chains (Fig. VI.7). This is important, for example, in relation to blood groups, which are determined by sugar molecules on the surface of blood cells (Box 4).

Box 4   Blood groups

Cells, including the blood cells, have various carbohydrate structures (chains, usually branched, of linked sugars) on their surfaces. These are potentially highly immunogenic. Normally in our own bodies they would not be so because they would be recognised as 'self', i.e. belonging to the body. However, since the introduction of blood transfusion, it has become apparent that there are genetic differences in the surfaces of the red blood cells of different individuals. These are recognised and classified in terms of blood groups. The best known is the ABO blood group classification. In this classification, a person may be A, B, AB or O. All four categories carry a structure on their cells' surfaces known as the H antigen. This consists of five linked sugars. The gene responsible for the difference between A, B, AB and O encodes an enzyme that can attach a sixth sugar unit, creating a branched structure. In Group A individuals, this enzyme selects a sugar called N-acetylgalactosamine. Group B people have a different version of the enzyme that attaches galactose instead. For most genes we have one copy from each parent, and AB individuals have inherited one copy of the A enzyme and one copy of B from their two parents. Finally, in group O people, the enzyme is either missing or inactive, so neither sugar is added and these people just carry the unadorned H antigen. If you are an AB, your immune system will recognise both A and B as 'self', and this means that you can safely receive blood from all four blood groups without raising an immune reaction. However, you can only safely donate blood to other ABs. At the other end of the scale, Group Os have never encountered either the A or the B sugar structures and treat them as non-self. Therefore, A, B or AB bloods will all raise an immune response and Os can only safely receive blood from other Os. A group A person will react against B or AB but can receive blood of Group A or Group O. They can donate blood to other Group As or to ABs. See if you can work out the pattern for Group Bs.

Similar considerations, only more complicated, apply in **organ transplantation**. Many different groups of surface antigens have to be matched, which is why it is so hard to find a suitable donor. A perfect match is seldom possible and hence transplant surgery is usually accompanied and followed by the use of immunosuppressive drugs to prevent rejection and destruction of the new organ.

Self-test MCQs on Chemistry VI

1 Taking account of aldose and ketose and asymmetric carbon, there are three different triose (three-carbon) sugars. On the same basis, how many pentose (five-carbon) sugars are there?

   (a) 7

   (b) 9

   (c) 12

   (d) 18

2 For a pentose sugar, such as ribose, how many different positions are there in its molecule through which it can be attached to another sugar to make a disaccharide?

   (a) 5

   (b) 4

   (c) 2

   (d) 1

TOPIC 11

Carbohydrates: sugars and polysaccharides in metabolism

Dietary carbohydrates

Carbohydrates are one of our two main dietary energy sources (the other being fat). We take in polysaccharides, especially **starch**, in bread, pasta, rice, potatoes, etc. We also take in disaccharides such as sucrose (table sugar) (C1 of glucose linked to C2 of fructose) and lactose (important for babies as the main sugar in milk) (C1 of galactose linked to C4 of glucose). We have separate enzymes to convert all these into monosaccharide sugars by splitting the glycosidic links, and more enzymes that interconvert different monosaccharides so that, instead of lots of separate breakdown pathways, we funnel everything into one pathway, the glucose breakdown pathway.

Starch and glycogen

Starch is a mixture of two types of polysaccharide molecules: amylose is a long chain of hundreds of glucose units all linked in the same way, C1 to C4 – a 1–4 linkage (Fig. 11.1); the other component is amylopectin, which also has strings of 1–4-linked glucose units but every so often uses one of the C6 positions to sprout a branch. If we join a C1 of a new glucose unit onto C6 of a glucose that is already in the 1–4 chain, then the new glucose has its C4 —OH free and can be the base of a new branch chain of 1–4-linked units. Depending on the source of the starch, these branches occur about every 30 units or so, and this gives rise to a huge molecule shaped rather like a tree (Fig. 11.2).

Pain-Free Biochemistry   Paul C. Engel
© 2009 John Wiley & Sons, Ltd

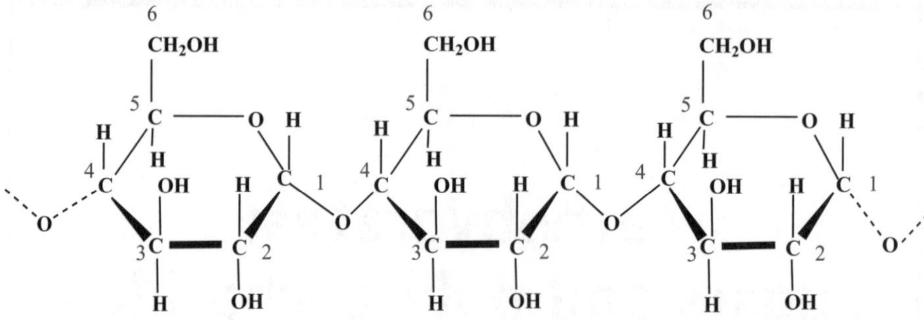

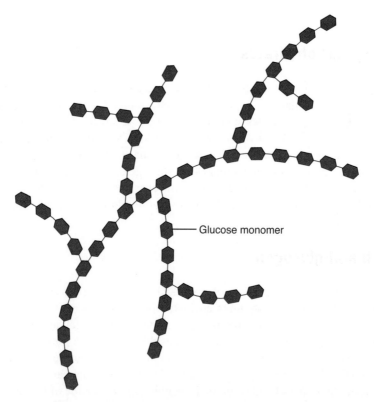

Figure 11.1 1,4-glycosidic links.

Figure 11.2 Branched structure of amylopectin (in starch) and of glycogen. Reproduced from Tortora and Derrickson (2009) *Principles of Anatomy and Physiology*, 12th edn, Wiley International Student Version, New York. © 2009 John Wiley & Sons. Reprinted with permission of John Wiley & Sons, Inc.

Starch is a plant polysaccharide that we encounter in our food, but we in fact manufacture a very similar substance ourselves: in our liver and in our muscles we have our own carbohydrate energy store, **glycogen**. This is closely similar to plant amylopectin, with a branched, tree-like molecule, the only difference being that glycogen is more highly branched.

Enzymatic breakdown of polysaccharide chains

We break down polysaccharides in our bodies in two different ways in our digestive system and inside our own cells. In our digestive system, the most important thing is to break the food down into molecules that are small enough and soluble enough to be taken up across the mucosal membranes of our alimentary tract. We therefore do the simplest thing, which is to split the glycosidic bonds with water, i.e. to 'hydrolyse'. The complex tree-like molecules of amylopectin need several enzymes to break all the links, but the most important one is **amylase** (in the mouth and in the small intestine), which splits 1–4 glucose links and so can break down amylose and also the long branches of amylopectin to release glucose for take-up. (Note the typical enzyme name, adding the ending 'ase' onto most of the name of the compound it works on.) Strictly speaking, this way of breaking down polysaccharides by hydrolysis is a bit energetically wasteful, and we shall see later that inside the body we do it in a more thrifty way.

The intestinal juice also contains enzymes to break down individual disaccharides. The one to break down lactose is very active in babies, as might be expected, but less so in adults. Its low activity in many Oriental people is responsible for **lactose intolerance** – milk products cause problems.

A glucose polymer that cannot be digested

There is one other carbohydrate that we take in very large quantities, **cellulose**. This is another polysaccharide, entirely made up of glucose units, 1–4 linked, and it is the material of plant cell walls, so that if we eat vegetables and fruit we are taking in huge amounts of cellulose. However, unlike cows, we cannot use it. How can this be? We have enzymes to break down the 1–4 links in starch. What is different? At this point, we need to return to glucose cyclisation discussed at the end of Chemistry VI. When we close the ring, the C atom that was the)C=O carbon is now carrying an H and an OH instead. In fact, its four bonds are each linking it with a different atom or group, and so it has become another asymmetric carbon – with two possibilities. The easiest way to see this is to look again at Fig. VI.6 and realise that the −OH at C1 could be either above or below the ring. These are spatially distinct. We said earlier that the ring and straight chain are rapidly interconvertible features of one chemical compound, glucose. So are the two ring forms not likewise interconvertible? Yes, indeed they are, but only so long as the −OH remains free.

As soon as it is involved in a glycosidic link, it is locked into one form or the other. Starch has one form (alpha linkage) and cellulose the other (beta), and enzymes will discriminate absolutely between the two. So we cannot digest cellulose. Even cows cheat because it is not their enzymes that break down the cellulose from grass, but the tanks of bacteria they carry with them!

Note that the fact that we do not make use of glucose from cellulose does not mean that cellulose is a useless or undesirable component of our diet. Dieticians speak of 'roughage', and the presence of this undigested bulk helps to maintain regularity and provide nourishment for the bacterial population of the colon.

Self-test MCQs on Topic 11

1 Human digestive juices do not break down cellulose. Why is this?

(a) Cellulose is not formed by 1–4 linkages between its sugar units.

(b) Cellulose is a monosaccharide sugar and does not need breaking down.

(c) Cellulose is not composed of glucose units.

(d) Cellulose has the wrong kind of 1–4 linkages between its sugar units.

2 Maltose from the breakdown of starch, lactose from milk and sucrose from fruits and sugarcane can all be broken down in our small intestine. Why is this?

(a) Because all of them have alpha 1–4 linkages.

(b) Because there is a separate specialised enzyme for each.

(c) Because all of them are based on glucose.

(d) Because amylase accepts a wide variety of different types of sugar unit.

The central role of the liver

Once we absorb glucose in the intestine, the hepatic portal vein, draining the intestinal area, carries it to the liver. The liver is a metabolic command centre, which allocates rations to the other tissues of the body, and in the case of glucose it has a key role (Topic 43) in maintaining the general blood glucose level constant at about 5 mM (see Chemistry III for definitions of chemical concentrations). Many tissues can use various energy sources interchangeably according to availability, but some, notably brain and other nervous tissue and red blood cells, only use glucose and will suffer if the blood glucose level drops. The liver has two tricks up its sleeve to maintain glucose levels during fasting. One is to use its extensive stores of glycogen. The second (see Topics 23 and 26) is to convert other compounds (e.g. amino acids) into glucose. In extreme starvation, we can draw on our own muscle mass as a last-resort store of amino acids to turn into glucose.

Depending on the metabolic situation, liver has the option of putting newly arrived glucose straight out into the bloodstream or of laying it down as glycogen 'for a rainy day'. As we shall see later, these metabolic choices are guided by hormone action (Topic 43).

Glucose breakdown and ATP

Assuming glucose is going to be used immediately as an energy source, how does this process work? The objective is to produce the 'energy currency' molecule, ATP (from ADP and phosphate) (see Topic 9). Simply asking ADP and phosphate to join up will not work, even with the help of an enzyme, because of the equilibrium issues looked at earlier (see Chemistry IV and Topic 9). It is precisely *because* this is an

Pain-Free Biochemistry   Paul C. Engel

(a)

(b)

Figure 12.1 In (a) the little man on the left thinks he is going to give his friend a surprise, sending him flying up into the air, but he is the one in for a surprise and a sore bottom. The energy as he hits the ground is just dissipated and lost into the surroundings. In (b) he has worked out that if you want to get energy from one process into another you have to have a way of directing it. The see-saw ensures that this time his evil plan is going to work!

uphill reaction that ATP is so useful to us once we have got it. It contains a lot of locked-up chemical energy that can be harnessed to drive biological processes when it breaks down to ADP. Conversely, therefore, making it requires an input of energy. It is useless, however, to have another reaction that releases the energy and then hope that the energy will somehow know where it is required (Fig. 12.1a). This would be like dropping a letter in the street and expecting it to find its own way to the right house and post box. The energy has to be *delivered* (Fig. 12.1b). So the trick is going to be to lead metabolism through a series of reactions that will make compounds that are 'happy' to hand on a phosphate to ADP in an energetically favourable reaction (i.e. one that will work!) and to provide a direct link between the driving and driven reactions (see Topic 9). We shall see how this is done in the sections that follow.

The metabolic pathways that break down glucose may seem needlessly, even perversely, complicated, with Nature conspiring to make the student's life difficult. If you find yourself in that frame of mind, it is worth returning to the previous paragraph, which explains why pathways have to be so cunningly constructed to enable one process to drive another. There is not a lot of point in trying to remember metabolic pathways without understanding what they are trying to achieve!

Self-test MCQs on Topic 12

1 Glycogen is which of the following?

  (a) our major dietary source of carbohydrate

  (b) laid down to meet the energy needs of the liver

  (c) laid down by the liver to meet the needs of other tissues.

2 The significance of glucose for energy metabolism is

  (a) that its catabolism is routed via compounds that will react with ADP to form ATP

  (b) that the glucose molecule undergoes a metabolic conversion into ATP and CO_2

  (c) that the oxidation of glucose releases heat and this can drive formation of ATP. Which?

TOPIC 13

Breakdown of sugar: glycolysis

Initial investment: sugar phosphates

Once glucose has entered the cell (see Topics 43 and 44), the first reaction comes as a surprise. We use a molecule of ATP to convert glucose into glucose 6-phosphate (G6P) (and ADP) (Appendix 5). This seems to be going in the wrong direction! We are using up ATP at this point instead of making it. Next we have an enzyme that converts the aldehyde sugar G6P into the ketone sugar, fructose 6-phosphate (F6P), and now we spend a *second* ATP to add a phosphate to the other end of the sugar molecule, making fructose 1,6-bisphosphate (Fig. 13.1). (The numbers 1 and 6 in these structures tell us which of the six carbon atoms have the phosphate groups attached.) We are now two ATPs down, but have to think of this as an investment phase, setting things up for the harvest ahead (Fig. 13.2).

Harvest reactions

Next we have a reaction that chops the six-carbon sugar phosphate in the middle, producing two three-carbon sugar phosphates (Fig. 13.3). One is the phosphate derivative of dihydroxyacetone (which we met in Chemistry VI) and the other is the corresponding derivative of D-glyceraldehyde, which we also met. (Note: It is pointless to burden you with the names of all the enzymes, but it is very important to realise that each and every one of these reaction steps has its own specific enzyme and that the metabolic pathway simply could not operate without them all.) The ingenious feature next is that there is an enzyme that interconverts the aldehyde and ketone triose sugars. This means that instead of having to have a separate

Pain-Free Biochemistry   Paul C. Engel
© 2009 John Wiley & Sons, Ltd

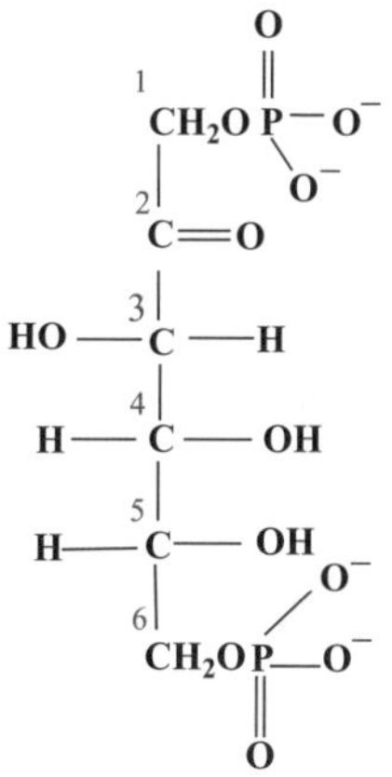

$$
\begin{array}{c}
\overset{1}{C}H_2O\,\overset{\displaystyle O}{\underset{O^-}{\overset{\|}{P}}}\text{--}O^- \\
\overset{2}{C}=O \\
HO\text{--}\overset{3}{C}\text{--}H \\
H\text{--}\overset{4}{C}\text{--}OH \\
H\text{--}\overset{5}{C}\text{--}OH \\
\overset{6}{C}H_2O\,\overset{O}{\underset{\displaystyle O}{P\text{--}O^-}}
\end{array}
$$

D- fructose 1,6-bisphosphate

Figure 13.1   A sugar phosphate: fructose 1,6-bisphosphate.

set of enzymes to deal separately with each of them, we just have one set, to deal with glyceraldehyde 3-phosphate, and all the dihydroxyacetone phosphate can be metabolised through the same route.

The next step is one of the most important steps in metabolism to understand because it encompasses what energy metabolism is all about. It is the first oxidation in the pathway and it oxidises the aldehyde (—CHO) group of glyceraldehyde 3-phosphate using the coenzyme, NAD^+ (see Appendix 6), which is reduced to NADH in the process. Normally, a chemist might expect an aldehyde to be oxidised to a carboxylic acid (—COOH) (this is a group of organic compounds we met in Chemistry III). If all that the reaction accomplished was to oxidise the aldehyde in this way, it would no doubt gallop along as a very favourable reaction, releasing

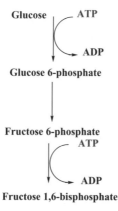

Figure 13.2   The 'investment' phase of glycolysis.

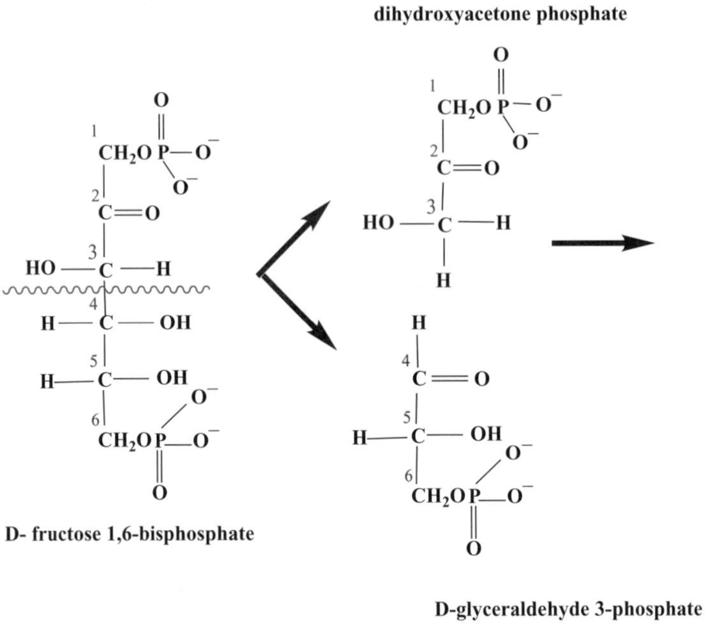

dihydroxyacetone phosphate

Figure 13.3 Splitting a 6-carbon sugar into two 3-carbon sugars.

energy as heat. This, however, is not what the cell needs to happen. Instead, the enzyme that is provided diverts the reaction along a different track so that, in addition to reacting with NAD⁺, the three-carbon fragment is forced to react with free phosphate ions floating around in solution, producing 1,3-diphosphoglyceric acid (instead of just 3-phosphoglyceric acid) (Fig. 13.4). Remember that the long-term objective is to convert ADP and inorganic phosphate to ATP. This is the point in the sequence where the phosphate has hopped on board waiting now for a chance to link up with ADP.

It does not need to wait long because the very next step provides an enzyme that plucks off the phosphate, transferring it to ADP to make ATP. Unlike the (theoretical) reaction of simple free phosphate ions with ADP, this is now an energetically favourable reaction, so that, instead of wasting all the energy at the aldehyde oxidation stage, we have siphoned a good portion of it off to drive ATP formation.

If you think back to our cartoon of the little man dropping through the air (Fig. 12.1), the idea of oxidising —CHO to —COOH and releasing the heat is similar to the man minus the seesaw; bringing the phosphate into the oxidation reaction is like bringing in the seesaw because what it does is to provide a direct link in the form of a common intermediate, a single substance, 1,3-diphosphoglyceric acid, which is made in the driving reaction and used in the driven reaction.

At this point, we get back an ATP for *each* triose phosphate fragment, i.e. two per original glucose. Bearing in mind that we invested two to start with, higher up the

glyceraldehyde 3-phosphate   phosphate

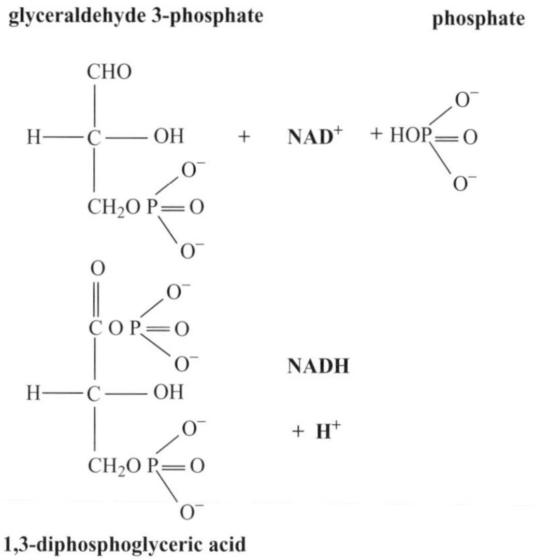

Figure 13.4   Oxidising triose phosphate (glyceraldehyde 3-phosphate) and bringing in extra phosphate.

sequence, we have now broken even, but this is not the end. The next two enzymes accomplish a little chemical shuffling, moving the phosphate onto the middle carbon position and then removing water to make a compound called phosphoenolpyruvate, very appropriately abbreviated and easier to remember as PEP. PEP has another phosphate ripe for plucking, and the next enzyme transfers this phosphate too to ADP, making ATP and leaving behind pyruvate. The score is now 4–2! We are in credit, with a net gain of two ATPs formed from ADP and free phosphate for each glucose molecule (Fig. 13.5).

Anaerobic vs aerobic metabolism

Two ATPs per glucose could be enough to live on, and the truth is that it is exactly what some cells and tissues do live on – red blood cells, for example. But notice that we have not used oxygen at all yet. This is **anaerobic** metabolism, occurring in the **cytosol** of the cell. In order to extract more usable energy in the form of ATP we need to use oxygen. For that we need the oxygen itself, a good blood supply to capture the oxygen in the lungs and carry it to the tissue, and also **mitochondria**, the sub-cellular particles (organelles) that contain the machinery to use the oxygen. These are not always available. Red blood cells carry oxygen and they *are* the blood supply, but they have no mitochondria. The lens of the eye is living tissue but would not be transparent if it were criss-crossed by blood vessels. In a sprinter's muscle fibres, there is a blood supply and there are mitochondria, but the oxygen cannot

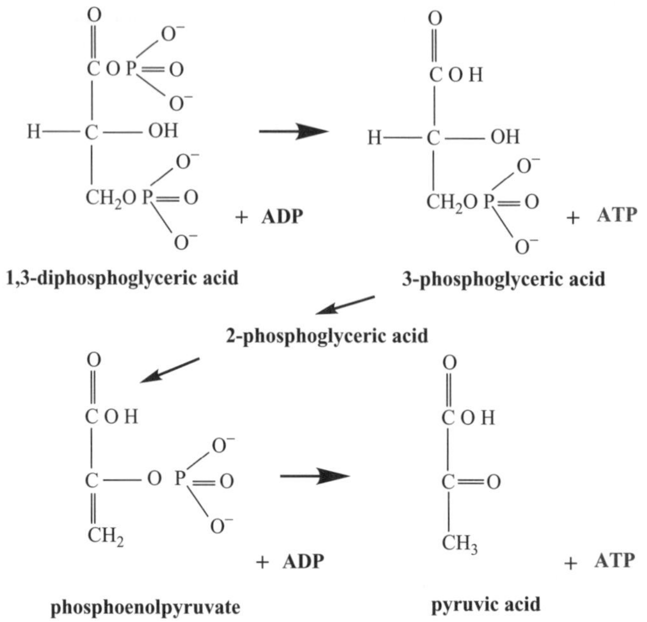

Figure 13.5 The 'harvest' or dividend phase of glycolysis. These reactions give back more ATP than was expended in the 'investment' phase.

get there fast enough if he is winning the 100 metres. So in all these situations the tissue has to rely on anaerobic metabolism.

Recycling the cofactor

There remains a problem – NADH, made from NAD^+ in the process of oxidising glyceraldehyde 3-phosphate (Fig. 13.4). As soon as all of the cell's small pool of NAD^+ has been converted into NADH we are stuck (see Box 3). Nothing can get past the glyceraldehyde 3-phosphate step without some NAD^+ to carry out the oxidation. If we were operating aerobically, oxygen would reoxidise all the reduced coenzyme molecules, but without oxygen we have to use something else. The cell has a neat solution. We have just produced pyruvic acid (pyruvate), $CH_3COCOOH$. The carbonyl $\rangle C{=}O$ can be reduced to $\rangle CHOH$, and this makes lactic acid, $CH_3CHOHCOOH$ (the sour taste in sour milk and yoghurt). The enzyme lactate dehydrogenase (LDH) catalyses this reaction, regenerating NAD^+ so that glucose breakdown can continue and give us more ATP. In rapid exercise, it is the accumulation of lactic acid in the muscles that causes pain:

$$\text{pyruvic acid} + \text{NADH} + H^+ \quad \leftrightarrow \quad \text{lactic acid} + NAD^+$$
$$CH_3COCOOH \qquad\qquad\qquad\qquad CH_3CHOHCOOH$$

Overall outcome

The net result of this process of anaerobic glycolysis is as follows:

$$1 \text{ glucose} + 2 \text{ ADP} + 2 \text{ phosphate} \rightarrow 2 \text{ lactic acid} + 2 \text{ ATP}$$

Note that no oxygen is involved as it is not available, that we have a net conversion of ADP and phosphate into ATP, and that the coenzyme NAD^+ does not appear in the equation since it has been recycled.

Self-test MCQs on Topic 13

1 Lactic acid is produced under anaerobic conditions

(a) to keep the pH value from rising too high

(b) to recycle NADH to NAD^+

(c) because lactic acid can be converted to CO_2 without oxygen

(d) to drive formation of ATP from ADP.

2 A major purpose of glycolysis is to achieve net conversion of ADP and inorganic phosphate (phosphate ions) into ATP. At what point in the pathway does the inorganic phosphate get incorporated?

(a) Formation of G6P.

(b) Formation of PEP.

(c) Splitting of fructose 1,6-bisphosphate.

(d) Oxidation of glyceraldehyde 3-phosphate.

3 Dihydroxyacetone phosphate is converted into glyceraldehyde 3-phosphate for which one of the following reasons?

(a) Because a molecule of ATP is formed in the reaction.

(b) Because dihydroxyacetone phosphate is toxic if allowed to accumulate.

(c) Because there is only one set of enzymes for dealing with oxidation of triose phosphate.

(d) Because this allows recycling of cofactor.

TOPIC 14

Aerobic oxidation of pyruvate: Krebs cycle

Conversion of pyruvate into acetyl CoA

As mentioned in Topic 9, we trap most of our energy (as ATP) by oxidising our foodstuffs to CO_2 and water. At the lactate stage, we have not accomplished any net oxidation at all. There has been one oxidation to make NADH, but we have had to 'give it back', using the half-processed foodstuff itself to reconvert NADH to NAD^+ so that glycolysis can continue. With oxygen, mitochondria and a good blood supply, things are very different (this means that different parts of the same body can be processing glucose differently at the same time – while the sprinter's muscles are making lactate, the brain, thinking about the finishing line, is oxidising glucose completely to CO_2 and water).

Under **aerobic** conditions pyruvate is free to enter the mitochondria, where it meets pyruvate dehydrogenase (PDH). This is often also referred to as the PDH complex because it is really a single enzyme but a mini-assembly line with three different enzyme activities working in harness. Together they oxidise pyruvate using NAD^+, remove its **acetyl group**, –COOH, as CO_2 and leave the other two carbon atoms attached to a new player, 'coenzyme A', making acetyl CoA, one of the most central and important metabolic intermediates in our cells. (CoA is the abbreviation for coenzyme A; Appendix 7.) In text-book chemical reactions you will often see acetyl CoA written as $CH_3COSCoA$ and CoA itself written as CoASH instead of just CoA. This is because CoA is not a chemical structure, just a name, and the SH highlights the chemical 'business end' of this cofactor molecule, a so-called sulphydryl or thiol group, i.e. a sulphur atom, S, with a hydrogen attached.

$$CH_3COCOOH + NAD^+ + CoASH \rightarrow CH_3COSCoA + CO_2 + NADH + H^+$$

Pain-Free Biochemistry Paul C. Engel
© 2009 John Wiley & Sons, Ltd

We have already mentioned two vitamin-derived cofactors involved in the PDH reaction (CoA and NAD^+), but in addition *three* others are involved – thiamine pyrophosphate (TPP) from vitamin B1, FAD from riboflavin (vitamin B2) and lipoic acid, so the PDH step depends on five different vitamins and would fail if any of them were missing.

Oxidation of acetyl CoA

Acetyl CoA feeds into a remarkable biochemical sequence, variously known as the tricarboxylic acid (TCA) cycle, the citric acid cycle or, after its discoverer, the **Krebs cycle** (Fig. 14.1). It is called a cycle because it does indeed end up where it started, at a compound called oxaloacetate. Oxaloacetate combines with acetyl CoA, releasing the CoA and forming citric acid (so named because plentiful in citrus fruit juices). A sequence of seven further reactions takes citrate back to oxaloacetate. What has happened is that the two carbons in the acetyl group of acetyl CoA have been taken up (into citrate) and two carbons have been separately released as CO_2 in the reactions of the cycle. If we tracked individual carbon atoms like ringed birds (as we can with radioactive isotopes – see Chemistry X), we would find that the cycle does not release precisely the same two carbon atoms that entered, but that need not concern us. In net chemical terms, it does not matter where the atoms came from. At the end of the cycle, the *amounts* of all the **TCA cycle** intermediates are unchanged, and in effect an acetyl group has been oxidised to $2CO_2$ (see Appendix 8).

If something has been oxidised, something must have been reduced. In spite of what we said about aerobic metabolism, it is not oxygen that has been reduced at this stage. Instead, the TCA cycle has four separate reactions where it uses oxidation cofactors. In three cases it is once again NAD^+ being converted into NADH; in the other one it is a different coenzyme, FAD, converted into $FADH_2$. Why is this? Why not use oxygen if we have got it on hand? This is because to do so would release all the energy we want to trap. We are holding onto as much of the chemical energy as we can because in the next major process of **oxidative phosphorylation**, carried out by the **respiratory chain (or 'electron transport chain')**, the mitochondria are able to couple the reoxidation of NADH and $FADH_2$ by oxygen directly to the formation of ATP from ADP and phosphate (see Topic 15).

One further important detail of the TCA cycle is that, as well as making CO_2 and a lot of reducing power as NADH and $FADH_2$, it also in effect produces one molecule of ATP directly. One of the cycle's reactions converts GDP and phosphate into GTP. GDP/GTP are chemically very similar to ADP/ATP, and in energy terms GTP and ATP are equivalent, much like, say French euros and German euros. There is an enzyme that will bring about that equivalence, converting GTP plus ADP into GDP plus ATP.

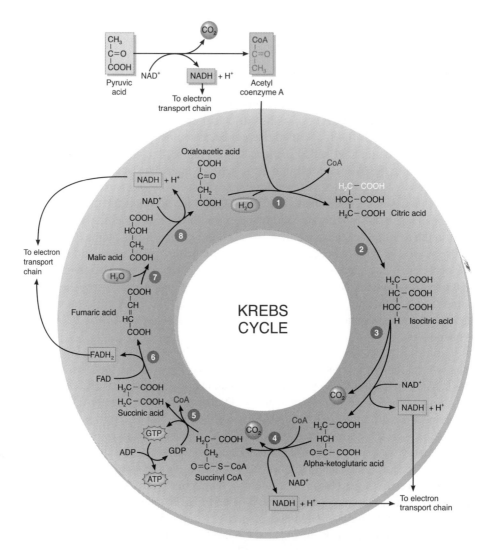

Figure 14.1 The Krebs tricarboxylic acid (TCA) cycle. Reproduced from Tortora and Derrickson (2009) *Principles of Anatomy and Physiology*, 12th edn, Wiley International Student Version, New York. © 2009 John Wiley & Sons. Reprinted with permission of John Wiley & Sons, Inc.

Self-test MCQs on Topic 14

1 The operation of the Krebs cycle achieves which one of the following?

(a) Recycling NADH to NAD^+.

(b) Conversion of CoA into acetyl CoA.

(c) Conversion of lactic acid into pyruvic acid.

(d) Conversion of acetyl groups into CO_2.

(e) All of these.

2 What is the effect of glycolysis on the concentration of the compounds (e.g. citric acid) that make up the Krebs cycle?

(a) Their concentrations increase by exactly twice the concentration of the glucose used.

(b) Their concentrations decrease as a result of CO_2 production.

(c) Their concentrations do not change.

3 The enzymatic conversion of pyruvic acid to acetyl CoA normally involves five vitamin-derived cofactors. If two if these five were unavailable over a prolonged period would you expect

(a) that the reaction would be unable to proceed

(b) that the body would synthesise the missing cofactors

(c) that the reaction would go at roughly 40% of its normal rate.

TOPIC 15

Respiratory chain, oxidative phosphorylation and overall ATP yields

The role of the mitochondrion

Wherever they occur in the cell, the reactions we have considered thus far all occur in free solution. Dissolved chemical substances find their way onto the surface of dissolved enzyme molecules which are likewise floating around in free solution, and the products drift away to find the next enzyme. The process that forms the majority of our ATP under aerobic conditions is rather different. The reducing equivalents from NADH and $FADH_2$, as we saw in Topic 10, are treated like a relay baton that must be passed from runner to runner and at all costs not dropped. In this case, the relay race is set up for a definite purpose and the 'runners' are a set of mitochondrial proteins that, to avoid any mistake, are lined up adjacent to one another and fixed in place in the inner mitochondrial membrane (Fig. 15.1). We shall consider in Topic 17 exactly what a membrane is, but for now it is a layer or sheet that separates one cell compartment from another. Very important for the present purpose, normally if there is a membrane there is an inside and an outside, i.e. the membrane encloses a sealed bag.

The respiratory chain and proton gradients

The sequence of relay runners, or carriers as they are actually called, is known as the respiratory chain and includes a set of proteins called the **cytochromes**. The

Pain-Free Biochemistry Paul C. Engel
© 2009 John Wiley & Sons, Ltd

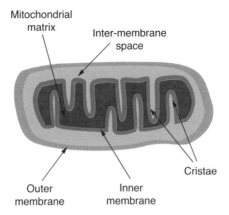

Figure 15.1 The mitochondrion: schematic diagram showing the inner membrane where most of aerobic ATP formation occurs and the matrix which is the site of fatty acid β-oxidation, the Krebs cycle, etc. Each typical cell contains many mitochondria.

'chrome' in their name refers to their strong colour, which derives from the fact that they all carry around a tightly attached cofactor. This one, haem, is a cofactor we can make for ourselves without any vitamins. In the middle of the haem structure, there is an atom of iron, Fe, and it is important for the function of haem that iron is an element, like copper (Chemistry V), that can change its valency. Iron can be ferr*ous*, with a valency of 2 (Fe^{2+}), but can be oxidised by losing an electron to become ferr*ic*, with a valency of three (Fe^{3+}).

The carriers of the respiratory chain are there to successively reduce one another in sequence until the 'reducing equivalents' are finally passed on to oxygen. However, if we consider the various carriers we can see that they divide up into two types. One group can be termed 'hydrogen carriers' because they become reduced or oxidised by gaining or losing a pair of hydrogen atoms:

$$XH_2 \rightarrow X + 2H$$

The other type, as we have just mentioned, are oxidised or reduced by change of valency and are therefore described as 'electron carriers'. If a hydrogen carrier reduces an electron carrier, removing the electrons to carry out the reduction leaves behind H^+:

$$H \rightarrow H^+ + e \text{ (e being an electron)}$$

The reverse is of course also true if an electron carrier reduces a hydrogen carrier. The physical arrangement of the carriers, along and across the membrane, then means that as they successively oxidise/reduce one another they also accomplish a pumping action that moves H^+ ions across the membrane (Fig. 15.2) (because of where each oxidation/reduction takes place). The pumping of protons, as shown in

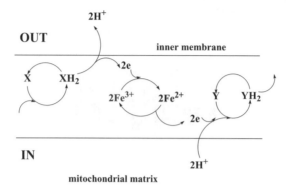

Figure 15.2 Electron carriers and hydrogen carriers: by alternating two types of oxidation/reduction the electron transport chain can pump H$^+$ across the membrane. This creates a gradient which can be used to drive ATP formation.

Fig. 15.2, happens at three stages along the electron transport chain, and so it has been generalised in the diagram. X is a hydrogen carrier and in its reduced form it has 2 H atoms that it can pass on. If what is waiting to receive them is an electron carrier, e.g. Fe^{3+}, then each H can split into a proton (H$^+$) and an electron. Each Fe^{3+} can pick up an electron, but this means the proton is left over. When the electron carrier in turn comes to reduce a hydrogen carrier further along the chain, it has to pick up a proton to go with each electron it passes on (H$^+$ + 1 electron gives us back an H atom). The arrangement cross the membrane as shown means that protons are picked up on one side of the membrane and discharged on the other. At the bottom end of the chain copper (Cu^{2+}) takes over as the electron carrier instead of Fe^{3+}.

Left to its own devices this machinery would make one side of the membrane more and more acidic (low pH) and the other more and more alkaline (high pH). What in fact happens is that there is a second assembly of membrane proteins, the so-called membrane ATPase. On its own this would simply squander ATP, using water to split ATP into ADP and phosphate, hence its misleading name. But this reaction also is linked to the movement of H$^+$ across the membrane. If the membrane does not leak, the gradient of H$^+$ created by the respiratory chain is able to drive the ATPase *backwards* so that it makes ATP instead of splitting it.

Coupled and uncoupled electron transport

The coupling of these two processes is not unlike the clutch in a car, which couples the energy of the combustion engine to the driving of the wheels and can drive the car up a hill it would otherwise rather run down. In a car it is possible to slip the clutch, so that the engine races and not much else happens. In the same way, it is possible to uncouple mitochondria totally or partially. This happens to some extent in some fevers. It is also put to use in 'brown fat', which is very important in hibernating animals but also in human babies. The common link is production of heat instead of ATP. We have talked hitherto of heat production as a waste of energy, but in winter it might be just what you need. There are various chemicals that are well known to be **uncoupling agents**, one being dinitrophenol, considered at one time as a slimming agent but very toxic!

The net reaction

There are still two important questions we have not mentioned. First, what happens to the reducing equivalents in the end – who receives the 'baton'? The answer is that at the 'bottom end' of the chain is a protein assembly called cytochrome oxidase, and the last part of its name tells you that it catalyses the handover to O_2, the oxygen breathed in through our lungs and carried to wherever it is needed by the bloodstream. This final reaction produces water, H_2O. The second question is about numbers – reoxidation of NADH and $FADH_2$ via the respiratory chain forms ATP from ADP and phosphate. But how much? For many years, biochemists agreed that the number was 3 for NADH and 2 for $FADH_2$. This was based on the best available experiments on mitochondria supplied with different substrates and on everyone's assumption that the number was bound to be a simple digit. More recently, however, a much better understanding of exactly how the ATPase works in molecular detail has indicated that the true numbers are closer to 2.5 and 1.5.

Overall ATP yields

We are now in a position to recap and work out the yields of ATP at the various stages of glucose metabolism. We saw that in anaerobic glycolysis there is a net yield of two ATPs per glucose molecule. That was based on the assumption that the NADH produced by oxidising glyceraldehyde 3-phosphate would be recycled by reducing pyruvic acid to lactic acid. Under aerobic conditions this does not need to happen, and the NADH can be reoxidised instead by making use of the mitochondrial respiratory chain. If we use the recent figure of 2.5 ATPs per NADH reoxidised, this means 5 ATPs per glucose (two molecules of triose phosphate produced from one glucose). Next, as we saw in Topic 14, the pyruvate can now be oxidised to acetyl CoA, again producing NADH. This offers another five ATPs per glucose. The complete oxidation of the acetyl group of acetyl CoA to CO_2 and water via the Krebs cycle produces three NADH, one $FADH_2$ and one GTP. In terms of ATP, this is equivalent to $3 \times 2.5 + 1 \times 1.5 + 1 = 10$ ATPs. However, this is only for one acetyl CoA and again there are two per glucose, and so the ATP yield from this phase of the process is $2 \times 10 = 20$ ATPs per glucose. Finally, we have still got the original two ATPs from the conversion of triose phosphate to pyruvate. This makes a net yield of 32 ATP molecules made from ADP and phosphate for every glucose molecule oxidised (Table 15.1). This is 16 times more ATP than harvested under anaerobic conditions.

That last striking number, 16, means that if an active tissue has not got an adequate source of oxygen, there needs to be a very good and immediately available source of glucose, as we shall discuss in Topic 16.

Table 15.1 ATP yield per glucose molecule

Direct formation of ATP in glycolysis	2 ATP
NADH from glyceraldehyde 3P dehydrogenase reaction	$2 \times 2.5 = 5$ ATP
NADH from pyruvate dehydrogenase reaction	$2 \times 2.5 = 5$ ATP
From TCA cycle – oxidation of 2 acetyl CoA molecules	$2 \times 10 = 20$ ATP
Total	32 ATP

Self-test MCQs on Topic 15

1 Which one of the following is true about the respiratory chain?

(a) It operates to make ATP in the mitochondria of aerobic cells.

(b) It provides the link between the lungs and other tissues of the body.

(c) It converts acetyl groups in acetyl CoA into CO_2 and water.

(d) It is a line of cytochrome molecules passing oxygen one to another in the red cell.

2 Oxygen-driven formation of ATP requires which of the following?

(a) ADP, NADH and $FADH_2$, phosphate.

(b) ADP, NADH or $FADH_2$, phosphate.

(c) Glucose, NADH and $FADH_2$, phosphate.

(d) Glucose, NADH or $FADH_2$, phosphate.

3 The net yield of ATP from aerobic metabolism of glucose is which of the following?

(a) 32 g of ATP per gram of glucose.

(b) 32 molecules of ATP per gram of glucose.

(c) 32 g of ATP per molecule of glucose.

(d) 32 molecules of ATP per molecule of glucose.

4 When the aerobic formation of ATP is complete, which of the following is true about the NADH involved?

(a) It has been oxidised to CO_2 and water.

(b) It has been reoxidised to NAD^+.

(c) It has been converted into ATP.

(d) It has been pumped across the mitochondrial membrane.

TOPIC 16

Mobilising the carbohydrate store: glycogenolysis

The advantage of glycogen in muscle

Returning to our sprinter from Topic 13, whose muscles have to do the 100 metres powered by anaerobic glycolysis, we should revisit glycogen. In rapid and sudden exercise, it is not only the oxygen supply that is inadequate; glucose cannot be supplied fast enough from the bloodstream, and so the muscle lays down its own local intra-cellular supply. As a special first line of defence, muscle has a certain amount of **creatine phosphate**, laid down in the good times, which can pass on its phosphate to convert ADP into ATP when there is a sudden demand. However, the main energy store in skeletal muscle is glycogen, and the glucose units have to be separated off the tree-like molecules in order to enter glycolysis. Back in Topic 11 we noted that splitting glucose units apart with water (hydrolysis) was energetically wasteful. In fact, in both muscle and liver this is achieved, not by amylase (Topic 11) and hydrolysis but by a different enzyme, glycogen phosphorylase. This attacks the sugar–sugar glycosidic links with phosphate rather than water, releasing glucose 1-phosphate molecules instead of glucose. A second enzyme quickly juggles the molecule, transferring the phosphate from C1 to C6 so that we have G6P. G6P is on the main high road of glycolysis (Topic 13), but we have also managed to bypass the initial ATP-utilising step from glucose to G6P. This means that our net ATP yield per glucose unit from glycolysis to make lactic acid is not $4 - 2 = 2$ but $4 - 1 = 3$. It is still a far cry from the yields of aerobic metabolism, and admittedly we will have had to pay up front (in ATP) in the good times in order to lay down the glycogen in the first place. Nevertheless, 3 is 50% better than 2, and when you are running from the tiger you will not ask too many questions about the cost of survival (Fig. 16.1)!

Pain-Free Biochemistry   Paul C. Engel
© 2009 John Wiley & Sons, Ltd

Figure 16.1   A lucky escape – thanks to glycogen!

A different role in liver

As already mentioned in Topic 12, the other major site of glycogen deposition and mobilisation is the liver. Unlike the muscle, which essentially is looking after itself, the liver, as usual, is taking care of everyone else and specifically is maintaining the blood glucose level, either in emergency situations or in fasting (see Topic 43). Note, however, that in this case we discard the advantage of having the glucose already phosphorylated (G1P or G6P). The readily transported form of the sugar is glucose rather than G6P, and so the liver uses its enzyme machinery to convert G6P into glucose for export (see also Topic 26 for more detailed consideration of this enzyme step).

Self-test MCQs on Topic 16

1 An important advantage of glycogen in muscle is which of the following?

(a) It is rapidly converted into glucose for immediate use.

(b) Its breakdown to sugar units forms ATP directly.

(c) Its breakdown yields ready-phosphorylated glucose, saving ATP.

(d) It is a better form than glucose to transport carbohydrate from the liver.

2 An important advantage of glycogen in liver is which of the following?

(a) It can rapidly meet the liver's heavy energy requirements.

(b) It can be broken down on the spot to G6P and then to glucose for export into the blood.

(c) Its proximity to the intestine means that it can rapidly be transported to be broken down by intestinal amylase.

CHEMISTRY VII

Alcohols, esters, glycerol, fatty acids and triglycerides

Alcohols and esters

In chemistry, if you take a simple hydrocarbon (carbon skeleton with only hydrogens attached) and replace one of the Hs with an —OH, a hydroxyl group, you have an alcohol. In Chemistry VI we met two simple ones: methanol and ethanol. We have also met carboxylic acids, compounds with a —COOH group, in Chemistry III.

It is also possible chemically to bring about a reaction between the —COOH of a carboxylic acid and the —OH of an alcohol, eliminating water and making a type of compound known as an **ester**. Methanol and acetic acid, for example, would give the ester methyl acetate.

$$CH_3OH + CH_3COOH \rightarrow CH_3COOCH_3$$

This might require fairly stiff conditions, e.g. cooking up in concentrated sulphuric acid. Living cells can make esters too, but they cannot duplicate the chemist's conditions. Cells would not take kindly to concentrated sulphuric acid and so instead they use enzyme-catalysed reactions to form esters under milder conditions.

Glycerol and glycerides

An important alcohol that forms esters in our body is **glycerol**. This has a molecule with only three carbon atoms, but all three carbons carry an —OH (Fig. VII.1a), so that glycerol can form an ester at each of these three positions. In our bodies what is most frequently 'esterified' at the three —OH positions of glycerol is **fatty acids**.

Pain-Free Biochemistry   Paul C. Engel
© 2009 John Wiley & Sons, Ltd

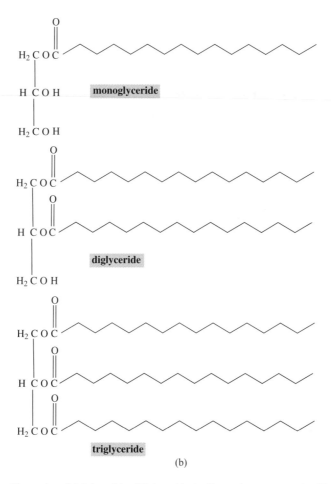

Figure VII.1   Glycerol and triglyceride. Triglyceride is the main component of body (and dietary) fat. In triglyceride (b) each of the —OH positions of glycerol (a) is occupied by a fatty acid attached by an ester linkage.

These are just the carboxylic acids we have already met, but instead of being 2 or 3 carbons long, they tend to be 16 or 18 carbons long; sometimes longer, sometimes shorter but nearly always an even number!

Depending on whether glycerol has one, two or three fatty acids attached, we have 'monoglycerides', 'diglycerides' or **triglycerides** (Fig. VII.1b). What we think of as 'fat', whether in our diet or carried around on our own bodies (where it can make up 20% of normal body weight and much more in the obese), is almost pure triglyceride. Another word you will encounter, used interchangeably as a term for fats, is lipids. This is a group of biomolecules that are important both in terms of cellular structure and for energy metabolism.

Self-test MCQs on Chemistry VII

1 A diglyceride has which of the following?

   (a) Two esterified fatty acid chains and one free —OH group.

   (b) One fatty acid chain and two esterified glycerol molecules.

   (c) Three free —OH groups and one free fatty acid.

   (d) Two linked molecules of glycerol.

2 Glycerol is which of the following?

   (a) A monoglyceride

   (b) A triglyceride

   (c) An ester

   (d) An alcohol

3 The chemical linkage between a fatty acid and glycerol is which of the following?

   (a) A peptide linkage

   (b) A glycosidic linkage

   (c) An ester linkage

CHEMISTRY VIII

Hydrophobic, hydrophilic and amphiphilic

Favourable and unfavourable interaction with water

To understand both the advantages and the problems associated with using fat in our bodies, we need to consider the physical properties of these (and similar) substances. If you take neat alcohol and tip it into water and shake it around a bit, it very quickly forms one clear layer. If you take solid sugar or salt you will need to shake or stir a little longer, but again the solid grains will disappear and you have one homogeneous, clear liquid. This is because these are all **hydrophilic**, i.e. water-loving substances. If you take vegetable oil and try and do the same thing, the suspension may briefly go opaque as the oil breaks up into tiny droplets, but the droplets do not dissolve. Instead, as each droplet touches another, they come together to make a bigger droplet, seemingly with obvious relief at escaping from the nasty water, until finally you have two completely separate clear layers again with the absolute minimum possible surface contact between the two substances. Vegetable oil is plant fat (triglyceride), and like all fats it is **hydrophobic**, water-hating and insoluble because of the long hydrocarbon chains of the fatty acids. Fats are therefore very compact in a water environment but by the same token insoluble and problematic to move around.

The hydrophobic effect governs many interactions in the cell, and not just for fats. In protein molecules, for example, hydrophobic interactions between some of the amino acid side-chains contribute to the way they fold up and then also to the way they interact with other structures in the cell.

Pain-Free Biochemistry   Paul C. Engel
© 2009 John Wiley & Sons, Ltd

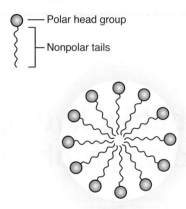

Figure VIII.1 A micelle. Reproduced from Pratt and Cornely (2004) *Essential Biochemistry*, John Wiley, New York. © 2004 John Wiley & Sons. Reprinted with permission of John Wiley & Sons, Inc.

Amphipathic molecules

This tendency for hydrophilic substances to club together, all paddling in the water, and for hydrophobic substances communally to shun the water is a powerful force in the cell. Let us now be slightly mischievous and make a molecule that is part hydrophilic and part hydrophobic in its structure. It could perhaps have a long hydrocarbon chain but carry a charged group at one end. What will such a compound do?

In fact, we are all familiar with such compounds (**amphipathic**, i.e. feeling both ways, both water-hating and water-loving at the same time) and most of us use them every day. These are detergents. In water, all the hydrocarbon chains bury themselves together, 'head down'. Escaping from contact with the water, they form little balls with a hydrophobic core. All the hydrophilic tails are on the outer surface of the ball in contact with the water (Fig. VIII.1). This way both parts of the molecule are satisfied. The remarkable properties of detergents arise because, unlike the droplets of vegetable oil, these little balls, **micelles** as they are called, are stable in suspension in water. Not only that, but because of their hydrophobic core they can dissolve, say, the grease on your plates, taking it safely into the core, out of contact with the water.

TOPIC 17

Phospholipids and membranes

Phospholipids

One of the most striking structural manifestations of the properties of amphipathic molecules is to be found in the **membranes** of our cells – both those that surround the cell and those that define the organelles inside the cell, e.g. mitochondria. These are composed of a special kind of fat (or 'lipid') that shows exactly the amphipathic properties discussed in the last section. These are phospholipids. Going back again to the structure of glycerol with its three —OHs, in phospholipids two of the positions are occupied by long-chain fatty acids, just as in triglycerides. However, the third is quite different (Fig. 17.1). Joined onto the third —OH is, first of all, a phosphate (see Appendix 5). Phosphate in the cell is always negatively charged, making this part of the molecule already hydrophilic, but a further piece is linked via the phosphate and this too is hydrophilic. There are various small chemical structures, e.g. **choline**, that can be attached, forming different classes of phospholipid, but they all add charge or else other hydrophilic groups such as —OH.

Although phospholipid molecules have a more complicated molecular structure than the detergent structure we considered in Chemistry VIII, the basic idea is the same: we now have two long hydrophobic bits, which, to start with, will tend to cling to one another, and a third bit, which, being highly hydrophilic, will want to avoid the other two, making a structure rather like a tuning fork (Fig. 17.2).

Lipid bilayer

Just like simpler detergents, phospholipids can form micelles, but perhaps because the hydrophilic and hydrophobic bits are more similar in length in a phospholipid

---

Pain-Free Biochemistry   Paul C. Engel
© 2009 John Wiley & Sons, Ltd

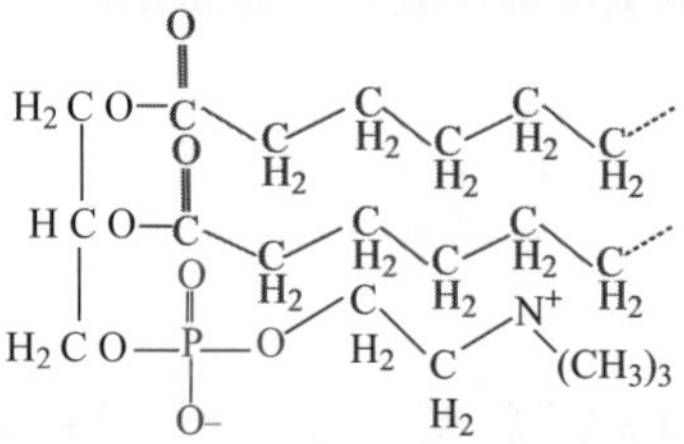

$$H_2 C O-\overset{\overset{O}{\|}}{C}$$

Figure 17.1 Structure of a phospholipid. Two of the three —OH positions of the glycerol molecule (black) are occupied by fatty acid chains (green), but in a phospholipid the third position is linked via negatively charged phosphate (red) to another polar component. In the particular case shown this is choline (blue) and the resulting phospholipid is lecithin (widely used as an emulsifier in foods). There are several other types of phospholipids with the choline replaced by other substances.

molecule, these structures can assemble in a different way that is crucial to the structure of our cells. Imagine grabbing a handful of sparklers, or incense sticks, making sure they all pointed the same way and tapping the bunch down on the table. With enough of them you would make a thick, even layer, with all the heads neatly lined up beside one another and likewise the tails. If we now imagine doing the same with hundreds of phospholipid molecules floating around in water, all the hydrophilic tails are in contact with the surrounding water, so they are 'happy'. The hydrophobic fatty acid chains that make up the head are all gathered together too, but there is just one problem: the layer that they make has an entire hydrophobic surface in contact with the water, and that is not good. Is there a solution that will allow this structure to work? All we have to do is put two of these layers back to back, and we now have a stable **lipid bilayer**. This is exactly what happens, and the lipid bilayer is the basic structure of all our membranes. They do contain other things, for example we saw in Topic 15 that the inner mitochondrial membrane contains respiratory chain carriers and also the protein assembly that makes ATP, but the basic structure supporting these extras is the (phospho)lipid bilayer (Fig. 17.3).

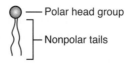

Figure 17.2 In a watery environment the hydrophobic fatty acid chains of a phospholipid will cling together, trying to escape the water. The hydrophilic tail will show the opposite behaviour.

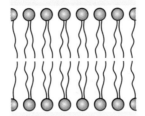

Figure 17.3 A lipid bilayer. Two sheets of phospholipids molecules come together so that all the fatty acid chains are entirely buried inside the sandwich and the polar, hydrophilic groups face the watery surroundings. Reproduced from Pratt and Cornely (2004) *Essential Biochemistry*, John Wiley, New York. © 2004 John Wiley & Sons. Reprinted with permission of John Wiley & Sons, Inc.

Self-test MCQs on Topic 17

1 In a lipid bilayer which of the following is true?

(a) The hydrophilic tails of one side of the bilayer are in contact with the hydrophilic tails of the other side.

(b) The hydrophobic chains of one side of the bilayer are in contact with the hydrophobic chains of the other side.

(c) Both a) and b) are true.

(d) The hydrophobic tails of one side of the bilayer are in contact with the hydrophobic chains of the other side.

2 A lipid bilayer tends to form which of the following?

(a) a ball

(b) a column

(c) a sheet

(d) a chain

Saturated and unsaturated

Double and triple bonds

It is difficult these days to escape headlines, labels and advertisements dwelling on saturated and unsaturated fats. Those with a bit of chemistry background will be familiar with the idea of saturation in the context of dissolving things, say salt in water. When you have dissolved as much as will dissolve, the solution is said to be saturated. The use of the term 'saturated' for fats, however, is absolutely unconnected with dissolving in water or in anything else.

If we go right back to carbon and its four valencies (Chemistry I), if two carbon atoms are joined together, each using up one valency, then both have three 'arms' left over. If we fill all of these with links to hydrogen atoms, we have the simple hydrocarbon ethane with the formula C_2H_6. However, our carbon atoms could have a more enthusiastic partnership and share not one but two valencies. This makes a carbon–carbon double bond. Now there are only four arms left over, and if again we fill them with hydrogens, we have C_2H_4, which is ethylene. (Plants use this as a ripening hormone.) We can even go one stage further and make a triple bond in C_2H_2, acetylene, used as a fuel in welding torches (Fig. IX.1). As soon as we start putting in double or triple bonds we have what a chemist would call an unsaturated organic compound – unsaturated with hydrogen. By chemical reaction one can put back the missing hydrogen, but if you start with ethane, of course you cannot because it is already saturated. It cannot take up any more hydrogen.

Double bonds in fatty acids

In the context of fats, the long fatty acid chains have the potential to be either saturated or unsaturated. A typical animal triglyceride is likely to have both saturated

Pain-Free Biochemistry   Paul C. Engel

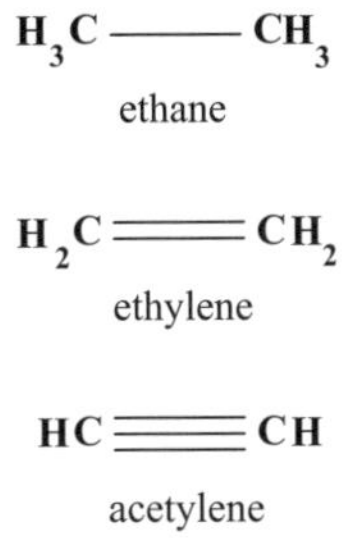

$$H_3C \text{———} CH_3$$
ethane

$$H_2C \text{====} CH_2$$
ethylene

$$HC \text{≡≡≡} CH$$
acetylene

Figure IX.1   Unsaturated hydrocarbons.

and unsaturated fatty acids esterified to the three —OH positions of its glycerol portion. Thus, a very common fatty acid in animal fat is stearic acid, the fully saturated 18-carbon fatty acid, but if we put in one double bond, we get oleic acid, also common in animal fat and abundant in *oli*ve oil. With two double bonds we have *lin*oleic acid, plentiful in *lin*seed oil, and three gives linol*en*ic acid, all with 18 carbons in the chain (Fig. IX.2).

What does this do to the fat? You will have noticed that we are referring to plant oils in the context of unsaturated fatty acids. We should also dwell on the fact that plant oils are oils, i.e. they are liquid at room temperature, e.g. cooking oils. This is not true for animal fats: butter, lard, beef dripping and so on are all solid at room temperature. Yet all of these substances are almost pure triglyceride.

This all makes physiological sense. All our lipids, including our membranes, would go rigid if they were even at 25°C, but we do not live at 25°C, we live at

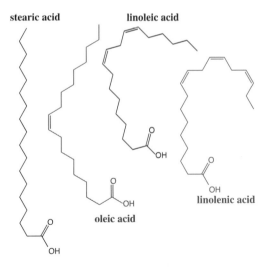

Figure IX.2   Unsaturated fatty acids.

37°C, at which temperature our lipids are liquid – tuck a packet of butter under your armpit if in doubt! That does not tell us how this works, however. The answer is quite simple. In a long molecule of saturated fatty acid like stearic acid, there is free rotation round all the single carbon–carbon bonds. This means that the molecules are free to twist and turn like a snake, which is what will happen when the temperature is warm enough for them to melt. It also means, however, that they can be pulled out into a straight zig-zag rod (Fig. IX.3). It is easy to line up large numbers of straight rods, and as they cool, say in a bowl in your fridge, this makes it easy for these saturated fatty acids to form a regular, rigid, solid structure. Once you put in a double bond, the unsaturated fatty acid cannot rotate around that particular point, it now has an inescapable kink and it cannot be pulled out into a simple, straight zig-zag rod any more. This makes it harder to line up these chains in a regular array, and as you heat them up, giving them more energy, they are more liable to escape and float around in the liquid state. This is why plant fats, rich in unsaturated fatty acids, have much lower melting points. Each type of organism has fats adapted to the temperature it lives at, so that the membranes especially can remain in a semiliquid state (Box 5).

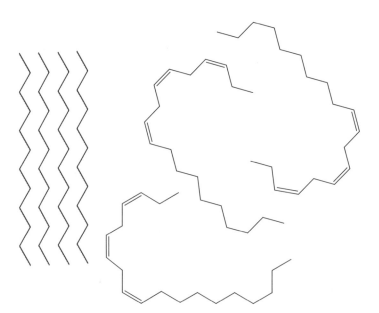

Figure IX.3   Kinked molecules are less easy to pack together. This is why saturated animal fats readily solidify (e.g. in your fridge) whereas unsaturated plant oils stay liquid.

Box 5   Saturated and unsaturated fatty acids in edible oils

The table below shows for a selection of significant dietary fats (triglycerides) of animal or plant origin the remarkable variation in the types of fatty acid attached to the glycerol. The two animal fats have only low amounts of the polyunsaturated fatty acids and do contain some cholesterol. None of the plant oils contains cholesterol. Soya and sunflower oils contain very high percentages of polyunsaturated fatty acids, but these figures are very much lower for palm and coconut oils and even for olive oil, a key component of the 'healthy Mediterranean diet'. Note, however, that olive oil is very low in saturated fatty acids, whereas coconut oil is 85% saturated! It is easy enough to think that 'fat is fat' and 'cooking oil is cooking oil', but these dramatic differences lie behind some of the dietary recommendations that are frequently made.

Fish oils, such as cod liver oil, are high in polyunsaturated fatty acids, when compared to, say, dairy fats, but still not as high as sunflower oil. However, they are seen as being especially healthy because an unusually high proportion of this polyunsaturated fatty acid is C20 with five double bonds (EPA) and C22 with six double bonds (DHA), so-called 'omega-3' fatty acids (see also Topic 49).

Oil	Saturated (no double bonds)	Monounsaturated (one double bond)	Polyunsaturated (two or more double bonds – 'essential')	Cholesterol
Lard	41	44	10	0.1
Butter	54	20	3	0.2
Coconut	85	7	2	0
Palm	45	42	8	0
Soya	15	23	57	0
Olive	14	70	11	0
Sunflower	12	0	63	0
Cod liver	20	43	23	0.5

Self-test MCQ on Chemistry IX

The food industry uses hydrogenated vegetable oil. Will the reaction with hydrogen

(a) increase the content of saturated fatty acids and make the oil less liquid?

(b) increase the content of saturated fatty acids and make the oil more liquid?

(c) increase the content of unsaturated fatty acids and make the oil less liquid?

(d) increase the content of unsaturated fatty acids and make the oil more liquid?

Fats as an energy source

Relative oxidation states of a fatty acid and a sugar

In a typical Western diet, apart from carbohydrate, fat is the most important source of energy, i.e. of chemical drive to form ATP. Just how important will depend on the individual, culture and so on, and nowadays also on the food fads of the moment.

We have seen that fat is both chemically and physically different to carbohydrate, and this confers some important advantages while at the same time raising logistic problems. First of all, bearing in mind that the objective is to carry out oxidation that will generate reduced cofactors to drive ATP formation (see Topics 14 and 15), how do the two foodstuffs compare? It is easy to do the arithmetic. If we take a typical long-chain fatty acid like stearic acid, $CH_3(CH_2)_{16}COOH$, we can work out how much oxygen is required to achieve the complete oxidation of this molecule to CO_2 and H_2O. Eighteen carbons will form 18 molecules of CO_2, containing $18 \times 2 = 36$ oxygen atoms. Stearic acid also has altogether 36 hydrogen atoms, which can form 18 H_2O molecules. These water molecules contain 18 oxygen atoms. This means that altogether the products contain 54 oxygen atoms after complete oxidation. The molecule of stearic acid already had two O atoms to get it on the way, so we need 52 extra oxygen atoms, or 26 O_2 molecules to achieve the oxidation.

We can now go through a similar calculation for glucose. Going back to Chemistry VI, we see that glucose is $CH_2OH(CHOH)_4CHO$. This adds up to $C_6H_{12}O_6$ (which tells us nothing about the structure but gives the total of all the atoms). Six carbons will give 6 CO_2 (12 O atoms) and 12 hydrogens will give 6 H_2O (6 O atoms). This makes a total of 18 Os, but this time we already have 6 to begin with in the glucose molecule, and so we only need to add a further 12, i.e. 6 O_2.

Now to do a fair comparison we need to bear in mind that stearic acid has 18 carbons and glucose has only 6. If, however, we oxidise 3 glucose molecules, we are also dealing with 18 carbons and will require $3 \times 6 = 18$ molecules of O_2. The thing to notice, therefore, is that 18 carbons worth of glucose requires 18 O_2 for its complete oxidation but 18 carbons worth of fatty acid requires 26 O_2. This is not surprising because even a glance at the two structures tells us that the glucose molecule is more highly oxidised to begin with, but the consequence is important – fat is more energy rich. In fact, if we do the comparison on a per unit weight basis instead of per carbon the comparison is even more striking. All those extra oxygen atoms are heavy! The relative molecular mass (see Chemistry II) of three glucose molecules is 540, while that of the stearic acid molecule is only 284, not much more than half.

Calorific values

These numbers underlie the fact, known for more than a century, that 1 g of pure fat is over twice as **calorific** as 1 g of pure carbohydrate (i.e. the complete combustion releases twice as much heat). The purpose of our metabolism is to ensure that a good proportion of that energy, instead of being released as heat, is used instead to drive ATP formation, and we may anticipate that also inside the body 1 g of fat will produce more ATP than 1 g of glucose.

There is, however, yet one more layer to the comparison: as we have discussed, fat is hydrophobic. In consequence, the fat stores in our bodies are indeed almost pure fat, containing very little water. Glycogen, the carbohydrate store, is hydrophilic and therefore will be highly hydrated in the cell, and consequently much heavier than pure, dry carbohydrate. Thus, in terms of real energy value per unit weight of stored material that we carry around with us, fat wins hands down.

Solubility and transport

There is, however, a penalty. Precisely because it is so compact and water resistant, fat is harder to mobilise and transport than freely water-soluble sugars. The body, therefore, has to resort to elaborate mechanisms to move fat round the body, and as we shall see these are crucially related to health issues, especially heart disease.

Self-test MCQs on Topic 18

1 As an energy source fat has the advantage that

(a) it is more readily soluble than other foodstuffs

(b) it is more chemically reactive than other foodstuffs;

(c) it is more calorific on a per gram basis than other foodstuffs;

(d) it needs less oxygen than other foodstuffs for complete oxidation.

2 Complete oxidation of palmitic acid, $CH_3(CH_2)_{14}COOH$, requires how many oxygen (O_2) molecules?

(a) 24

(b) 23

(c) 16

(d) 15.

TOPIC 19

Fats: digestion, transport, storage and mobilisation

Digestion of dietary fat

The rich potential for harnessing the energy of oxidation of fatty acids is realised in the mitochondria. This will be discussed in detail in Topic 20, but first we must consider how the fatty acids get there. Just as with carbohydrates, there are two issues to consider – fatty acids derived directly from our food and fatty acids derived from fat stores in 'adipose tissue'.

Although our food contains a mixture of various kinds of fat (lipid), we shall concentrate on triglyceride, much the most important quantitatively. The digestion of triglycerides occurs mainly in the small intestine. As discussed earlier, fats are not soluble in water, and therefore will not be in solution as the watery stomach contents enter the intestine. At this point they meet two other inputs, from the pancreas and from the gall bladder (**bile**). For fat digestion the bile plays a crucial role because the bile salts have a detergent action that breaks the fat up into a finely dispersed emulsion. The pancreatic juice contributes a cocktail of digestive enzymes, including a **lipase**, an enzyme that hydrolyses the triglycerides' ester bonds, releasing the two outer fatty acids from the glycerol. The lipase works at the surface of the fat droplets, and the emulsification action of the bile means that there is a vastly increased contact surface area for the lipase to work on. The main products are free fatty acids and 'monoglyceride', glycerol that still has one fatty acid chain attached at the middle —OH position (see Chemistry VII).

Pain-Free Biochemistry   Paul C. Engel
© 2009 John Wiley & Sons, Ltd

Intestinal absorption, processing and onward transport of fat breakdown products

Through the detergent action of the bile salts, the monoglycerides and fatty acids produced by lipase action are incorporated in micelles (Chemistry VIII), which are able to cross the brush border of the intestine into the mucosal cells.

Perhaps a little surprisingly, the mucosal cells proceed to reassemble triglyceride molecules from the pieces, and these are now repackaged with the help of a specific protein molecule, called B-48, to form **chylomicrons**. These stabilised food parcels are released from the cells and drained into the lymph. This eventually feeds them into the bloodstream via the thoracic duct.

The bloodstream will thus present triglyceride-rich chylomicrons to the various tissues. Imagine a cumbersome item of furniture: the van can get it up your street and your drive, but finally it has to be dismantled to get through the door. Trying to get triglyceride into the cell is a similar proposition, and so another lipase is present in the cell membranes of muscle cells and adipose tissue. This is able to unload fatty acids into those tissues by splitting the triglyceride molecules. Muscle will take up the fatty acids to use as an immediate energy source, whereas adipose tissue will take them up to store away and will yet again convert fatty acids back to triglyceride.

If there is a need to mobilise these fat reserves, then an intracellular lipase is activated and puts out free fatty acids into the blood. Once again there is a solubility problem, and these fatty acids are transported by carrier proteins in the 'albumin' fraction of the blood.

Multiple lipases

Note that we have invoked lipase action several times but we have referred each time not to lipase, but to *a* lipase. This is because it is a different lipase each time. They all carry out the same chemical job, but they have to work in different places under different conditions – the intestinal one is a secreted enzyme working free in the intestinal juice, whereas the lipase at the surface of adipose tissue is a membrane-bound enzyme. Similarly, the lipases of adipose tissues are tailored for their separate jobs by being structured to respond to different physiological regulatory signals. Thus, the lipase involved in fatty acid mobilisation, for instance, is known as 'hormone-sensitive lipase'.

Serum lipoproteins

This is not yet the full story, however, because we have not yet taken into account a set of lipoproteins that are prominent in the blood serum and are very important in relation to health, and also we have ignored one of the most important sites of fat metabolism, the liver. The lipoproteins, as their name implies, are partly protein and partly lipid, and they have a rather strange set of names that go back to a period when it was not at all clear what they all did. In the first instance, they were separated and recognised as distinct and different by the way they behaved when the serum was 'spun' at high speed in a **centrifuge**. This process subjects a sample to '*g* forces' many thousand times the normal gravitational field. In the case of these lipoproteins this separates them into separate bands in a centrifuge tube because they all have slightly different densities and therefore different tendencies to sink or float. Accordingly, they were named high-density (HDL), low-density (LDL), intermediate-density (IDL) and very low-density lipoproteins (VLDL) (IDL is intermediate between 'low' and 'very low', not between 'low' and 'high'). These practical descriptions have stuck even though they tell us nothing directly about the function of these proteins. They do, however, give us a hint about composition: triglyceride has a lower density than water (oil floats) and these four types of lipoprotein range in triglyceride content from about 5% for HDL up to about 60% for VLDL. They also show a gradation in size from VLDL, light but large, to HDL, dense but diminutive.

We can think of VLDL as a kind of food parcel prepared by the liver to send out to other tissues such as muscle. As well as the triglyceride, about 20% of the content of VLDL is either **cholesterol** (see Fig. II.6) or 'cholesterol ester' (meaning a compound in which a fatty acid is linked through its −COOH group to the −OH of cholesterol to make an ester; see Chemistry VII). The VLDLs are circulated round the body in the bloodstream, and they unload triglyceride into target tissues as they go. They gradually increase in density as this proceeds, and they become first IDL and eventually LDL. At this stage, the package has become a delivery messenger for cholesterol and cholesterol esters to peripheral tissues. HDL, containing a different set of proteins, has the opposite role as far as cholesterol is concerned: It ferries excess cholesterol to the liver where it may, amongst other things, be excreted in the bile, either directly or after conversion into bile salts such as cholate. (Cholesterol that has come out of solution in the gall bladder is the most common component of **gallstones**.) Because of these opposing functions in relation to cholesterol, LDL and HDL are frequently described by clinicians as 'bad cholesterol' and 'good cholesterol', respectively. This is because high levels of cholesterol are a primary cause of the deposition of **atherosclerotic plaques** in the coronary arteries and elsewhere, and a high HDL to LDL ratio will tend to keep the cholesterol level low.

All these lipoproteins can be thought of as rather elaborate micelles (see Chemistry VIII). All the non-polar, hydrophobic components are hidden in the core of the spherical particles, while the protein and phospholipids provide a hydrophilic outer layer that keeps the whole package stable in suspension in the watery environment of the blood serum. These, however, are 'intelligent' micelles because the proteins in these various lipoproteins are not there just for packing purposes. They also have quite specific functions in targeting the destinations and in unloading the cargo on arrival.

Self-test MCQs on Topic 19

1 Dietary fat is solubilised mainly by which of the following?

  (a) Chewing action in the mouth.

  (b) Mixing with acid in the stomach.

  (c) Detergent action in the small intestine.

  (d) Peristalsis in the colon.

2 Lipase action converts which of the following?

  (a) A fatty acid into a micelle.

  (b) Bile into bile salt.

  (c) Glycerol into glyceride.

  (d) Triglyceride into fatty acids and monoglyceride.

3 Chylomicrons are assembled in which of the following?

  (a) Intestinal lining.

  (b) Pancreas.

  (c) Thoracic duct.

  (d) Lymph.

4 Which of the following is true about HDL?

  (a) it is a form of cholesterol rendered harmless by the action of LDL.

  (b) it is a form of LDL made heavier by attachment of cholesterol.

  (c) it is a protein that ferries cholesterol to the liver.

  (d) it is a clinical danger sign associated with atherosclerosis.

Fats: oxidation of fatty acids

How do we break down the long hydrocarbon chains of fatty acids?

If we consider a long-chain fatty acid with 16, 18 or even 20 carbon atoms in its chain and think about converting it into CO_2 and water, there is an obvious question: Where do we start? At the end? If so, which end? In the middle? Do we break it up into pieces first? If so, what size of pieces? We have known the answer to all this, in outline at least, for a surprisingly long time – about 100 years. It was shown in dogs that fatty acids chemically tagged at one end, the end furthest from the $-COOH$, could still be oxidised. Almost everything was oxidised except for the end that was tagged, which was left behind as a sort of chemical stump. However, the nature of the stump varied according to whether the dog was given fatty acids with an even or an odd number of carbons. With even numbers, the stump still contained two of the original carbon atoms in the fatty acid chain; odd numbers gave a stump with only one of the carbons left (Appendix 9).

What does this mean? If we guess that the fatty acid molecules are somehow chopped into pieces exactly two carbons long, starting at the $-COOH$ end, this is exactly the result we should expect. In fact, it was proposed that fatty acids are metabolised by a process of 'β-oxidation'. Appendix 9 explains the numbering/labelling of fatty acid chains. In the older system, the carbon *next* to the $-COOH$ was designated the α (alpha) carbon, the next one along the β (beta) and so on. So the proposal was in fact that somehow the β CH_2 unit of the fatty acid was oxidised to a β CO, producing a molecule that could perhaps now be split, perhaps by water, perhaps releasing CH_3COOH, acetic acid. There was a lot of 'somehow' and 'perhaps' in the theory. Biochemists usually try to fill in the gaps and rapidly prove or disprove a metabolic theory by doing the experiments and finding the hypothetical intermediates in the pathway, i.e. the chemical stages on the way. That is how the pathway for glucose was worked out, but it did not seem to work for fatty acids – for nearly 50 years – even though the hypothesis was essentially right!

The role of CoA

One key ingredient had to be discovered first, and that was CoA (Appendix 7). We met CoA in the context of pyruvate oxidation and the TCA cycle. It turns out, however, that fatty acids go through β-oxidation, not as free fatty acids, but as the CoA derivatives of those fatty acids. Fatty acids are actually a bit sluggish chemically and attaching the carboxyl group to the S of CoA, just as in acetyl CoA, 'activates' them, making them more reactive. This had to be known in order to go looking for the enzymes, and even then it proved difficult to detect the metabolic intermediates. This is because there is only a limited 'pool' of CoA in the cell and that has to be shared out between all the intermediates – as we shall see there are a lot of them! So the intracellular concentration of any one intermediate is likely to be very small.

Transport to the site of oxidation

Now we are in a position to reveal what actually happens to a long-chain fatty acid on the way to total oxidation. If we start with stearic acid, the saturated 18-carbon fatty acid, and assume that it has just been unloaded into, say, a liver cell or a muscle cell, it has to make its way to a mitochondrion because that is where most* of the oxidation of fatty acids occurs.

Next, it is activated by an enzyme attached to the outer membrane of the mitochondrion, which attaches it to CoA, expending a molecule of ATP in the process.

$$RCOOH + CoASH + ATP \rightarrow RCOSCoA + AMP + \text{pyrophosphate}$$

This, in a sense, creates a problem because CoA carries several negative charges and therefore we have now made a molecule (stearoyl CoA) that is much too hydrophilic to be able to cross the double membrane of the mitochondrion without help. The solution lies in a specific carrier system for shuttling the fatty acid chains across the membrane barriers. It uses a molecule called **carnitine**, which, unlike CoA, carries no net charge. There is a system of three membrane-bound enzymes that (1)

---

*The word 'most' is there because some cells in certain physiological states have 'peroxisomes', another kind of organelle, which are distinct from mitochondria. Peroxisomes also use molecular oxygen to oxidise various substances, including fatty acids, but do so in a way that produces more heat and conserves less energy. This is beyond the scope of our book.

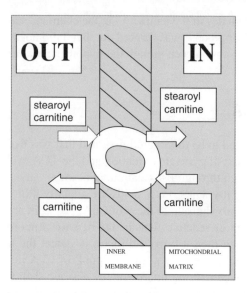

Figure 20.1   The carnitine shuttle. This brings fatty acyl chains into the mitochondrial matrix and because it exchanges acyl carnitine going in for carnitine going out, the carnitine supply doesn't run out in either cellular compartment.

transfer the fatty acid chain from the sulphur (S) atom of CoA to an oxygen atom on carnitine, (2) after the stearoyl carnitine has diffused across the outer membrane under its own steam, smuggle it across the inner membrane by a kind of revolving door mechanism that swaps stearoyl carnitine on the outside for empty carnitine on the inside (this ensures that there is always enough carnitine on the outside to do its job); and (3) finally at the inner surface of the inner membrane transfer the fatty acid back from carnitine to CoA (Fig. 20.1). This may seem an elaborate way of getting back, chemically speaking, to where you started, but it is part of the price for being able to divide the cell into separate, specialised compartments, each with its own border and immigration controls!

The reactions of the β-oxidation spiral

Now at last we can get down to the oxidation. There is a set of four reactions that is repeated again and again on a fatty acid chain, which gets gradually smaller with each cycle. Two of these four reaction steps are oxidations. As we saw in Topic 10, we can oxidise either by adding oxygen or by removing hydrogen. In β-oxidation, the first enzyme uses FAD to remove a hydrogen each from the α and β carbon atoms, creating a carbon–carbon double bond. Note that this gives us a reduced cofactor $FADH_2$. The next enzyme destroys the double bond by adding a water molecule across it. The α carbon goes back to its CH_2 state, but the β carbon is now

CHOH. The third enzyme catalyses a second oxidation, using NAD^+ to remove two more hydrogen atoms, this time from the CHOH, converting that grouping to $>C=O$.

Overall, this achieves exactly what was proposed a century ago on the basis of the experiments with dogs: conversion of the β-carbon position from CH_2 into CO. How do we now go down a step from 18 carbons to 16 carbons? Originally we said 'perhaps hydrolysis', and hydrolysis between the α and β carbons would give back a 16-carbon fatty acid. If so, however, the very next thing would be to activate this back to the acyl-CoA state, and, as we have seen, this costs ATP. Evolution is thrifty, and, instead, craftily, we use CoA itself instead of water as the attacking molecule to split the 18-carbon chain. Thus, the fourth enzyme in our sequence takes an 18-carbon β-ketoacyl CoA molecule, splits off the end two carbons with their own CoA to give a familiar molecule, acetyl CoA, and gives back the shortened, C16 chain ready primed with the new CoA attached. Smart! This means that we only need one activation step to go round all the multiple cycles of the β-oxidation sequence.

The next step is for the C16 acyl CoA to go through the same set of four reactions, and then C14 does the same and so on until, after eight trips round the circuit, the 18-carbon fatty acid chain has been broken down to nine molecules of acetyl CoA (Fig. 20.2).

ATP yield from β-oxidation

Let us now consider what this does for us in terms of energy trapping. Each of the eight cycles gives us one $FADH_2$ and one NADH. From Topic 15 we know that this is worth four ATPs overall. However, each cycle also gives us an acetyl CoA, and this can now enter the Krebs cycle and give three NADH, one $FADH_2$ and one GTP, adding up to ten ATPs. In other words, the oxidation of each C2 unit by the β-oxidation pathway gives rise to conversion of 14 ADPs into ATP. Adding up, we have $8 \times 14 = 112$ for the eight turns of the spiral pathway, and the last split leaves a ninth molecule of acetyl CoA without any further oxidation, and these last two carbons are worth another ten ATPs (The reaction yields two molecules of acetyl CoA, but we already counted one of them among the eight). This adds up to a total of 122 ATP molecules. However, we have to pay our debts and must not forget that we used ATP to activate the fatty acid in the first place. In fact, strictly we should subtract 2, rather than 1, because this activation step converts ATP not into ADP but into AMP. Since our energy currency relates to the interconversion of ATP and ADP, conversion into AMP counts as two units spent.

Overall, then, we have a total net yield of 120 ATPs! This compares with 96 ATPs produced through the aerobic oxidation of three glucose molecules (also containing a total of 18 carbon atoms). The sugar thus delivers 5.33 ATPs per carbon atom, whereas the fatty acid yields 6.66 ATPs per carbon.

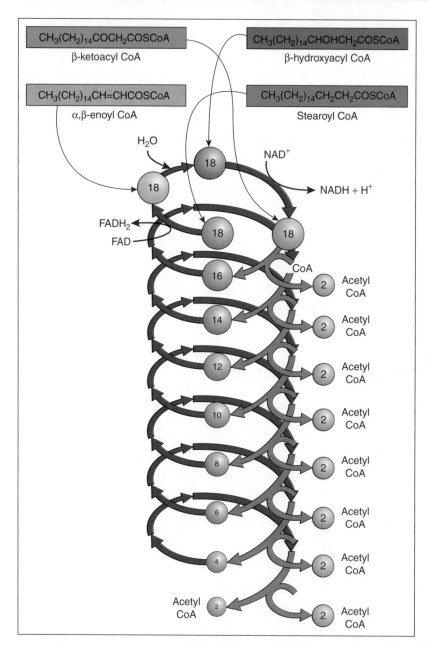

Figure 20.2 The descending spiral pathway of fatty acid β-oxidation. Starting with an 18-carbon fatty acyl CoA (C18), the same pattern of four consecutive reactions is repeated eight times – oxidation by FAD, hydration, oxidation by NAD⁺ and then a split using another CoA molecule to give acetyl CoA and leaving the fatty acyl CoA, now two carbons shorter, to re-enter the sequence. The C18 starting material thus yields nine molecules of acetyl CoA.

Box 6   Beta-oxidation and sudden infant death

'Cot death' or sudden infant death syndrome (SIDS) is extremely distressing and also, by definition, perplexing. The label seems to imply that it is a defined illness, but in truth SIDS is an administrative label, something to put on a death certificate when the paediatrician and the pathologist do not know the answer. Gradually, however, real causes are being found, showing that SIDS is not a single condition but a collection of diverse causes of death. One of the twists is that, as soon as a cause is identified, in a sense it ceases to be a cause, since SIDS defines death from no known cause.

One such identified cause, recognised about 20 years ago, is a genetic deficiency in the β-oxidation pathway for fatty acid oxidation that is described in Topic 20. As we have seen, very long, 18-carbon chains are progressively broken by a set of four reactions repeated over and over. It might appear that this should require only four enzymes, one for each chemical reaction. However, that would be like expecting one pair of shoes to do service for an entire family of eight or nine children, from the grown-up eldest right down to the toddler! We therefore have a set of enzymes to cover the range, and, in the case of the first oxidation step, there are three, one for long fatty acyl-CoA substrates, one for medium chains and one for very short ones. What happens now when there is a genetic deficiency affecting one of these steps? There is a certain amount of overlap between the abilities of the different enzymes in the set, and, in the case of the medium-chain enzyme (MCAD, standing for medium-chain acyl-CoA dehydrogenase), both the long- and short-chain enzymes can take over some of the load if the medium-chain one cannot cope. What we have then is a situation where the baby can just about cope but has no safety margin and is therefore constantly and invisibly vulnerable. Nothing may go wrong unless and until there is an episode of fasting, caused perhaps by missing a feed, often because of a minor infection. Fasting automatically leads the infant to mobilise its fat reserves and start relying more heavily on fat oxidation. This is the straw that breaks the camel's back. The defective β-oxidation machinery can cope with a steady trickle but not a flood. Unable to meet energy requirements adequately via fatty acid oxidation, these children go into a hypoglycaemic (low blood sugar) crisis and can easily die. The first time this happens in a family it is entirely out of the blue. Once this has happened, however, and been correctly diagnosed, if the episode was not fatal then the problem can be avoided very easily by ensuring that the child has frequent feeding with a diet high in carbohydrate and low in fat. If the first episode has led to a cot death then the first diagnosis allows screening of any other children in the family and safe, simple management. In the case of such a genetic cause of cot death (see also Topic 36) there is likely to be a one in four chance that any other child of the

(Continued)

same parents will be affected too. Astonishing unawareness of this on the part of some coroners has led to severe and tragic miscarriages of justice in which mothers have been imprisoned for the assumed murder of their own children!

Although MCAD deficiency is not common, it is common enough for some health authorities to screen for it. In a situation where diagnosis allows simple treatment and failure to diagnose may lead to a fatality, the case for screening is obvious.

Self-test MCQs on Topic 20

1 Fatty acids are broken down metabolically by which of the following?

(a) Random cleavage into manageable fragments by lipase action.

(b) Systematic removal of one carbon unit at a time for conversion to CO_2.

(c) Removal of two-carbon units from the carboxyl end.

(d) Attachment via glycerol to CoA.

2 CoA does which of the following?

(a) Chemically activates fatty acids for oxidation.

(b) Solubilises them for transport into the mitochondrion.

(c) Receives pairs of hydrogen atoms from the β carbon.

(d) Transports essential but insoluble carnitine across the mitochondrial membrane.

3 Each turn of the β-oxidation spiral involves which of the following?

(a) One oxidative step (requires NAD^+).

(b) Two oxidative steps (requiring NAD^+ and FAD).

(c) Two oxidative steps (both requiring NAD^+).

(d) Four oxidative steps (all requiring NAD^+).

4 Which of the following is the true statement about carnitine?

(a) Carnitine is an essential amino acid.

(b) Carnitine is a key phospholipid in membranes and chylomicrons.

(c) Carnitine is an intermediate in the Krebs cycle.

(d) Carnitine helps long fatty acid chains to enter the mitochondrion.

TOPIC 21

Ketone bodies in health and disease

What are ketone bodies?

The oxidation of fatty acids occurs in a number of major tissues of the body, including skeletal and cardiac muscle, liver, and kidney. In liver, however, there are two alternative possibilities as the dwindling fatty acid chains reach the C4 stage. One, as we have seen in Topic 20, is for the C4 ketoacyl CoA ($CH_3COCH_2COSCoA$, acetoacetyl CoA) to be split into two molecules of acetyl CoA for further metabolism. However, the liver is rather like the catering manager for the whole body, making sure that all the other tissues are well fed! One of the several types of 'food parcel' it sends out is in the form of four-carbon 'ketone bodies'. This term covers acetoacetate (CH_3COCH_2COOH), β-hydroxybutyrate ($CH_3CHOHCH_2COOH$) and also one other compound, acetone, which we met in Chemistry VI. Acetone, in fact, being a ketone, gives the name to the group, but in a way, although useful for removing nail varnish, it is a metabolic irrelevance or mistake! It arises without any help from enzymes by a spontaneous chemical reaction of acetoacetate, which is not very chemically stable. Acetone is, however, of medical interest because its odour is very characteristic and before the days of insulin therapy it was a clinical indicator of uncontrolled diabetes, for reasons we shall come to in Topic 44.

Ketogenesis, a normal function of the liver

It is important first of all to recognise that **ketogenesis**, the synthesis of ketone bodies from fatty acids, is a constant process in the healthy body. The liver is specialised for their production, by virtue of possessing two enzymes that are missing in other

Pain-Free Biochemistry   Paul C. Engel
© 2009 John Wiley & Sons, Ltd

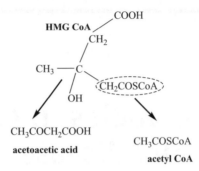

Figure 21.1 HMG CoA and ketone body production.

tissues, and two other tissues, the heart and the kidney, are similarly specialised to use them.

Since fatty acid β-oxidation is automatically going to be making β-hydroxybutyryl CoA and acetoacetyl CoA, it might seem pretty obvious that all you need to do is knock off the CoA, perhaps using a molecule of water to hydrolyse it off. However, the liver has a more devious route. Instead, it takes acetoacetyl CoA and acetyl CoA and joins them to make a new branched compound called HMG CoA. In the process, one of the two CoA molecules is released. In a second enzyme-catalysed reaction the HMG CoA is split to give us back acetyl CoA and leave acetoacetate without any attached CoA (Fig. 21.1)! If we look for a rationale for Nature adopting this apparently perversely complex route, this way of doing things makes the production of ketone bodies very sensitive to the concentration of acetyl CoA, and this makes good metabolic sense as we shall see.

A second important reason for making HMG CoA is totally unconnected with **ketone body** production. HMG CoA is also the starting point for making a number of other important biochemical compounds, including cholesterol and all the steroid hormones. The next step in making these compounds is the conversion of HMG CoA into mevalonic acid (no need to worry about structures here), which is catalysed by an enzyme called HMG CoA reductase. This enzyme is the target for one of the biggest-selling categories of drugs, the **statins**, which are prescribed, especially to middle-aged men, to slow down our natural production of cholesterol, since it is a contributor to the formation of atherosclerotic plaques in heart disease.

Getting back to ketone bodies, once the liver has made acetoacetate, it can be readily reduced to β-hydroxybutyrate. In fact, the two compounds exist together in an equilibrium balance. Both of them are highly soluble, pass through membranes without difficulty and are therefore put out into the bloodstream by the liver. For the same reason, they can readily enter kidney and heart, and, as mentioned above, these tissues have an enzyme specialisation, in that they possess an enzyme that takes succinyl CoA, a Krebs cycle compound, and swaps the CoA with acetoacetate so that we end up with succinate and acetoacetyl CoA. This is now all set to get straight back into energy metabolism as seen in the last section, by being split into

Table 21.1   Sites of ketone production and utilisation

Tissue	Liver	Kidney and heart	Other tissues, e.g. muscle
Ketone body production	Yes	No	No
Ketone body utilisation	No	Yes	No

two acetyl CoA molecules and metabolised via the Krebs cycle. The fact that the liver does not possess the enzyme to convert acetoacetate back to acetoacetyl CoA makes the ketone production machinery a bit like a one-way valve. The liver has no alternative but to deliver the goodies to the willing recipients, the kidneys and the heart (Table 21.1).

Ketosis

In spite of their entirely innocent and helpful role in normal metabolism, these well-meaning compounds have acquired a bad name in the clinical context. This is because ketosis, an excessively high level of ketone bodies, arises in two situations. The first is starvation or fasting or extreme exertion (e.g. prolonged labour in childbirth). The second is type 1 diabetes (see Topic 44). In both cases the reason is the same: unavailability of glucose forces tissues to metabolise fatty acids. Low levels of glucose will also tend to deplete the Krebs cycle carrier compounds, decreasing the capacity of the cycle. If acetyl CoA cannot be as readily oxidised via the Krebs cycle, it will tend to accumulate, and, as mentioned above, this now tends to switch more of the flow towards HMG CoA and ketone bodies. Ketosis is a sign that the system is struggling, and since the two acids, acetoacetic and β-hydroxybutyric, are quite strong acids, it tends also to challenge the pH balance. Therefore, in the event of ketosis arising from fasting, etc. (with low blood glucose), one would set up a glucose drip. This would *not* be appropriate in diabetic ketosis – the blood glucose in that situation is already too high!

Self-test MCQs on Topic 21

1 Which one of the following statements about ketone bodies is untrue?

   (a) Ketone bodies include acetoacetate, β-hydroxybutyrate and acetone.

   (b) Ketone bodies are constantly produced in the healthy state.

   (c) The main site of ketone body utilisation is the liver.

   (d) Ketosis refers to an excessive accumulation of ketone bodies, e.g. in diabetes.

2 Ketogenesis would be favoured by all the following except which?

    (a) A large throughput of glucose.

    (b) A large throughput of triglyceride.

    (c) Depletion of Krebs cycle intermediates.

    (d) Prolonged fasting.

3 Statins work by which of the following mechanisms?

    (a) They stimulate uptake of ketone bodies by the heart.

    (b) They block uptake heart's uptake of ketone bodies leading to excretion in urine.

    (c) They divert C2 units from ketogenesis to production of steroids.

    (d) They block production of cholesterol from HMG CoA.

Dietary fat: essential fatty acids

Unsaturated fatty acids in the diet

Many recommended diets nowadays attempt to cut back on fat intake. The very high intake of saturated fats is seen as a major cause of the very high levels of heart disease (atherosclerosis) in countries like Scotland and Ireland. So should we therefore aim to eliminate fat from our diet, and, if so, what should replace it?

The answer is emphatically 'No'! Unlike carbohydrates, fats contain components that we cannot make for ourselves from any other food source. These are the 'essential fatty acids' and they are essential in the same sense as vitamins – without them we become ill and ultimately cannot survive. They are all polyunsaturated fatty acids, i.e. they have several double bonds in their molecules. We mentioned some of these in Chemistry IX – linoleic and linolenic acids. As well as the actual number of double bonds in a fatty acids chain, two other things are important: the position(s) in the chain and the shape. If we start with the shape, there are in fact two possibilities for each double bond. If we imagine a double bond in the middle of the chain, then the two halves of the chain can both be either on the same side of the double bond ('*cis*') or on opposite sides ('*trans*') (Fig. 22.1). These two possibilities are not interchangeable because, as we saw in Chemistry IX, unlike single bonds, which can swivel, double bonds are rigid and allow no rotation. This has become an important diet/health issue. The double bonds in most natural polyunsaturated fatty acids are *cis*. In the manufacture of margarine and similar fats, food chemists have tried to turn plant oils into something with spreading properties more like those of semi-solid animal fats like butter. To do this, they use a process of chemical reduction (hydrogenation) to decrease the number of double bonds. This process, however, seems to result in products in which some of the remaining

Pain-Free Biochemistry   Paul C. Engel
© 2009 John Wiley & Sons, Ltd

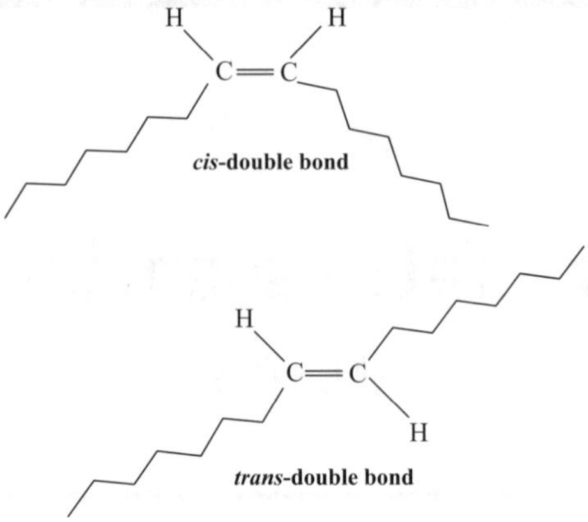

Figure 22.1 *Cis* and *trans* double bonds.

double bonds are now *trans*. These '*trans* fats' have got a very bad press, because they have every bit as bad an effect in terms of encouraging deposition of cholesterol, etc. as the original animal fats they were intended to replace. There is therefore a great deal of argument and campaigning over the fats used in baking, for example.

Turning to the question of where the double bonds are, our own enzymes are able to take a saturated C16 or C18 fatty acid and oxidise it to insert a double bond. We already know this can be done between carbons 2 and 3 in the process of β-oxidation, but that produces a transient metabolic intermediate. We are also able to insert a *cis* double bond between carbons 9 and 10 in long fatty acid chains. With a C18 chain this would convert stearic acid into oleic acid. This is a very commonly found fatty acid in human triglycerides. However, we have no enzymes that would allow us to put in a second double bond between carbons 12 and 13 to make linoleic acid, and therefore we have to get this through our diet. Linoleic acid is required so that we can make arachidonic acid, which has 20 carbons and 4 double bonds and is a precursor of an important class of signalling molecules, the eicosanoids, including prostaglandins, thromboxanes and leukotrienes. These are potent compounds involved in inflammation, pain, control of blood pressure, etc. (see Topic 49). Essential unsaturated fatty acids can be obtained from some plant oils and also from fish oils. So oil of evening primrose could help, but a nice grilled mackerel could be cheaper! (See Box 5.)

Self-test MCQs on Topic 22

1 Polyunsaturated fatty acids are of importance in nutrition because

  (a) they stiffen up plant oils to a spreadable consistency

  (b) they are a risk factor for heart disease

  (c) they are important metabolic precursors that we cannot make for ourselves

  (d) they assist the absorption of proteins.

2 Polyunsaturated fatty acids are plentiful in

  (a) all plant oils but not animal oils

  (b) in certain plant oils and certain animal oils

  (c) only fish oils

  (d) all plant oils and certain animal oils.

TOPIC 23

Protein and amino acid breakdown

Diverse foodstuffs converge on central catabolic pathways

At this stage of any biochemistry course, the student's head tends to be reeling as the complex sequences of compounds and reactions seem endlessly to pile up! It could be very much worse, however. If you consider the very different structures of a six-carbon sugar like glucose and of a long-chain fatty acid like stearic acid, it is remarkable that our metabolic pathways are so constructed that they converge on a single compound, acetyl CoA, which can then be processed via the Krebs cycle and the electron transport chain. This is all the more remarkable when one adds in the fact that protein breakdown can also feed into the same terminal pathway of oxidation, and particularly impressive when one looks closely at the building blocks of proteins, i.e. amino acids. Looking at different sugars, and even more so at different fatty acids, there is a degree of sameness. However, with proteins, the whole point is that the 20 different kinds of amino acids offer a wide variety of different structures, shapes and characters. To steer them all into the same metabolic funnel is quite a biochemical juggling feat!

Protein in the diet

First, however, we need to consider the overall status of proteins in relation to fats and carbohydrates. For most societies, protein is the expensive treat in the diet. Only in hunting communities, like traditional Inuit in the Arctic regions, would protein be abundant enough to be a major part of the calorie intake. Also we cannot

Pain-Free Biochemistry   Paul C. Engel
© 2009 John Wiley & Sons, Ltd

store protein in the same way as the other major foodstuffs. Excess fat can be laid down in adipose tissue. Excess carbohydrate can be either laid down as glycogen in muscle and liver or converted into fat. Excess amino acids can be built into muscle protein but, normally, in a well-fed adult, only under the influence of exercise and/or hormones. Muscle protein is indeed a large store of material, but it is one we only turn to in biochemical desperation, when we are starving. When the small glycogen store has gone and the large fat deposits are used up, there is little else to turn to, and so we have to keep ourselves going by becoming weaker! Posters from starving Africa should make it clear to us that muscle wasting is not the ideal answer.

We do need a steady input of protein because our own proteins are constantly turning over and we cannot be 100% efficient in recycling them. We are also constantly losing protein via our skin, our intestines, etc., and it has to be replaced. In growing children and in healing, e.g. after surgery, there is also a need for a net input to make new protein. Many of the amino acids can only be obtained by breaking down other protein, and so, without a steady input of high-quality protein, we inevitably become ill (see Topic 35 and Box 7).

Protein breakdown in the digestive tract

On the other hand, when we do eat protein, particularly in affluent Western societies, we often eat more of it than we strictly need. Since we cannot store the excess as protein, we scavenge the carbon skeletons of the amino acids instead.

To make dietary protein available to our bodies, it has to be broken down in the gut by digestion, just as fat and carbohydrates are. In our digestive tract, the food encounters a series of different enzymes that can split the peptide bonds that link one amino acid to another. Pepsin works in the acid contents of the stomach, while trypsin, chymotrypsin, elastase, carboxypeptidase and aminopeptidase all come in with the pancreatic juice and work in the alkaline environment of the small intestine. Some of these enzymes chop protein molecules in the middle, others nibble in from one or other end. Also some of them have particular preferences for the amino acids they like to cut at. It is not essential to know the details, but rather to realise that none of them would succeed on their own, while, working together as a team, they are able to break down most proteins into individual, separate amino acids. It helps if the protein is cooked because this 'denatures' the protein, unfolding its tight biological structure and opening it up to attack by the enzymes. The amino acids can be taken up through the intestinal lining and go into the bloodstream to be used wherever they are needed.

Zymogens

Enzymes that degrade proteins (known as **proteinases** or proteolytic enzymes) are potentially a danger to themselves since they *are* proteins. Not only that, but while they are sitting around in, say, a pancreatic cell waiting for action, they might chew up the cell that made them! Rather than run such risks, our cells produce these enzymes in inactive precursor forms that remain harmless until they are switched on, generally at the same time as they are introduced to their intended target. These enzyme precursors are known as zymogens. The process of switching on a zymogen often involves a cutting a peptide bond or removing a few amino acids, i.e. one proteinase acting on another.

Intracellular breakdown of proteins

A second source of amino acids for the pool of 'spare parts' is breakdown of proteins within our own cells. This has to be very carefully regulated, since uncontrolled **proteolysis** would wreak total destruction. Proteins that have passed their sell-by date or are no longer needed, and also foreign proteins, say from captured bacteria, are therefore targeted for destruction inside the cell and broken down by special intracellular protein breakdown systems. The detailed regulation of this breakdown is very complex, still being worked out and beyond our scope, but, briefly the systems include:

1 Breakdown of selected proteins by a structure called the **proteasome**. This is like a small machine for degrading proteins, containing an assemblage of protein-digesting enzyme activities. It only admits proteins that have been specifically labelled for breakdown by tagging with a small protein called **ubiquitin**.

2 Breakdown, particularly of ingested extracellular proteins, in the lysosome. **Lysosomes** are intracellular organelles with a very acidic content, rather like the cell's very own stomach! They contain a range of digestive enzymes including **cathepsins**, which degrade proteins.

3 Caspases. We have to recycle not only molecules but also whole cells, e.g. in growth and repair of organs. This involves a very finely regulated process of 'programmed cell death', also known as **apoptosis**, and the key step in switching on apoptosis is activation of a specific set of protein degrading enzymes called caspases. They are also involved in inflammatory processes and are therefore of great clinical interest.

4 Calpains. These are calcium-dependent proteolytic enzymes that are also involved in control of cellular processes.

Reabsorption of amino acids

Finally, there is a third source of amino acids, which is really more a way of plugging a leak! When the kidney filters our blood, initially small molecules like amino acids pass out into the liquid that will become the urine. However, as the liquid passes along the kidney tubule, the cells have specialised pump proteins (Topic 50) specifically there to recapture key valuable commodities, including amino acids.

Transamination

The circulating amino acids can, as we have said, be used for biosynthesis, i.e. for incorporation into new proteins, and this is covered in Topic 35. However, if they are surplus to requirement, the amino groups have to be removed in order to release the rest of the amino acid structures for further metabolic handling. Removing the $-NH_2$ converts an amino acid into a keto acid (or oxo acid) with a $\rangle C=O$ in place of the $-CHNH_2$. This is accomplished for many amino acids by **transaminases**. (Fig. 23.1).These enzymes use one of our B-vitamin-derived cofactors (pyridoxal phosphate from vitamin B6, pyridoxine), and they funnel all the amino groups onto one particular receiving keto acid, α-ketoglutarate (from the Krebs cycle), making the corresponding amino acid, glutamate. In Topic 24, we shall see how the nitrogen of the amino groups is further processed to get rid of it in the urine without poisoning us in the process!

Catabolism of the carbon skeletons of amino acids

Meanwhile, what happens to the keto acid that is left behind? In some cases, the keto acid can feed straight into our familiar central pathways. The amino acid alanine gives the keto acid pyruvate (see Topic 13). Likewise the amino acid aspartate gives

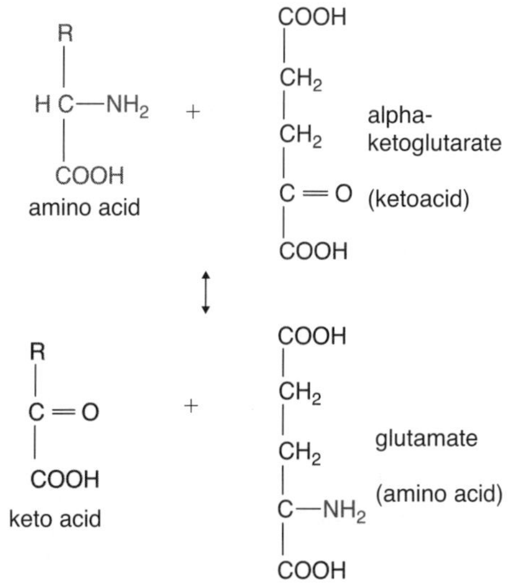

Figure 23.1   Transamination. This process moves the amino group from one keto acid to another. Glutamate and α-ketoglutarate are always one of the two pairs involved, so that glutamate serves as a clearing house for amino groups. The R in the two compounds on the left is not an unusual atom. It is the chemist's shorthand way of generalising – i.e. this could be any amino acid sidechain.

oxaloacetate, one of our Krebs cycle molecules. Others have to be further broken down, often by decarboxylation, removing the —COOH as CO_2. There are too many amino acids and too many separate routes for it to be worth looking at them in detail, but the striking thing is that nearly all these diverse structures end up broken down into one or more of a small handful of molecules – pyruvate, acetyl CoA, acetoacetyl CoA, fumarate, succinyl CoA, oxaloacetate and α-ketoglutarate. These are *all* either Krebs cycle molecules or else molecules that can feed into the Krebs cycle and so be oxidised via that route to CO_2 and water. *The Krebs cycle is thus the common pathway of catabolism for all the major foodstuffs, carbohydrate, fat and protein.*

Amino acid carbon skeletons can also be converted into ketone bodies and sugar

There is an alternative. Although we cannot store the amino acids themselves, we can use the fragments from their breakdown in ways other than immediate complete oxidation. Much of the processing of amino acids occurs in the liver, which does its best to look after the energy needs of the other tissues of the body. This becomes very important in starvation or prolonged fasting because our brain and nervous system require glucose as their energy source. If we have used up all our glycogen and have no food in our bellies, where can the glucose come from? If we are able to break down protein, then a look at the end-product fragments listed above re- veals that some of them (acetyl CoA and acetoacetyl CoA) can form ketone bodies, which will help meet the energy needs of heart and kidney. The other fragments, however, pyruvate and the Krebs cycle compounds, are all able to be converted into glucose (see Topic 26). Accordingly, amino acids can be classified as 'ketogenic', 'glucogenic' or both.

Box 7   Essential amino acids

From the standpoint of nutrition one always needs to think about the conse- quences of particular components being missing from the diet. For example, we now know that we need a supply of pentose sugars such as ribose to make the building blocks for deoxyribonucleic acid (DNA), ribonucleic acid (RNA) (see Topics 33–35) and also a variety of other important biochemicals such as ATP, NAD^+, FAD, etc. So what happens if there is no pentose in the diet? The answer is in Topic 28: if we have glucose or a source of glucose, then we can convert six-carbon sugar into five-carbon sugar. So we do not need to have pentose in our diet. What about amino acids? Twenty different ones are needed in order to make our proteins. What happens, therefore if there is, say, no alanine (Topic 4) in our diet? Alanine is readily made from pyruvate, and pyruvate arises in the metabolism of sugars, and so we do not need to take in alanine as such in our food. What about another amino acid such as tryptophan? This, the largest of the 20 protein amino acids, has a double ring structure that our cells cannot make for themselves. The only way to obtain it, therefore, is to eat it, relying ultimately on other organisms that do have the ability to synthesise it from scratch, and it is important to realise that many organisms that we regard as 'simpler' are nevertheless more metabolically versatile and independent than we are.

Since we cannot make tryptophan, we need to consider what happens if we cannot obtain it. Will our protein factories perhaps substitute a similar amino acid and keep going, or will they just skip one position? The answer is that everything will grind to a halt. The whole basis of the protein synthetic machinery is that it is incredibly specific and unerringly puts in the right amino acid (see Topic 35). Its job is not to be imaginative and resourceful, merely to carry out orders! However, it can only do so if provided with the necessary raw materials. So, if the right amino acid is not there, the process simply cannot proceed.

Is this just an biochemistry lecturer's trick question? After all, if we eat protein we should be taking in all 20 amino acids used to make them. That is true, and it works, *provided we eat a range of good quality proteins*, e.g. meat, fish, eggs, cheese, etc. However, if the diet relies very heavily on one particular protein one can run into trouble. How can this arise? For example, poor people in East Africa rely very heavily on maize as their staple diet. We might think of maize mainly as a source of carbohydrate, but they are relying on it for their protein as well. The problem is that instead of a wide selection of cellular proteins the predominant protein they are getting from the maize grains is one particular protein, zein, which the maize plant had laid down as a store with its own baby maize plant in mind, not us! This particular protein unfortunately contains *no* tryptophan, and so people in this part of the world may show symptoms of amino acid deficiency.

Tryptophan is therefore an essential amino acid, but it is not the only one. In fact, we can only make half of the 20 amino acids for ourselves. Lysine is another essential one that is liable to be deficient in some populations heavily dependent on grain crops, and one of the objectives of research to produce GM crops is to overcome problems such as this by genetically engineering codons for lysine into the grain protein.

Amino acid nutrition, however, is not just an issue for third-world countries. For various reasons people in more affluent societies restrict their diets, and vegetarian and vegan diets may contain insufficient quantities of 'first class' protein. This is for rather similar reasons to the maize problem. Vegetarians often rely heavily on beans for their supply of protein, and beans, like maize, contain large amounts of single storage proteins intended for the young plant. It is risky, therefore to rely too heavily on one kind of bean. Varied sources of protein are essential for health.

Self-test MCQs on Topic 23

1 Which one of the following statements about our dietary supply of amino acids is untrue?

(a) All proteins provide equally good sources of amino acids for human health.

(b) 'Essential' amino acids are those we cannot synthesise for ourselves.

(c) If one of the essential amino acids was missing we would be unable to make most of our proteins.

(d) Amino acids can be successfully supplied in our food intake either directly as amino acids or via protein breakdown.

2 Digestive breakdown of dietary proteins takes place

(a) mainly in the saliva (mouth and oesophagus)

(b) in the saliva and in the stomach

(c) in the stomach and the small intestine

(d) mainly in the colon.

3 Protein degrading enzymes

(a) break down proteins one amino acid at a time from the carboxyl (—COOH) end

(b) break down proteins two amino acids at a time from the amino (—NH$_2$) end

(c) break down proteins by cutting at random in the middle of the chain

(d) break down proteins by cutting from both ends and in the middle.

Shedding excess amino groups: urea cycle

Glutamate and deamination

As we have seen, in order to use amino acids as an energy source, we first have to strip off their amino groups, the NH_2 bits. Transamination funnels all the amino groups into glutamate, but glutamate is also an amino acid, so at first sight we might seem not to have moved very far forward. However, glutamate has its own special 'escape route' in the form of an enzyme, glutamate dehydrogenase, that removes the amino groups once and for all, not by transamination but *de*amination, which turns them into ammonia, NH_3:

$$\text{glutamate} + NAD^+ \leftrightarrow \alpha\text{-ketoglutarate} + NADH + NH_4^+$$

This small soluble molecule would be a very convenient end point if we were fish, because it would simply wash out through our gills. However, we are not aquatic animals*, and ammonia is a very toxic compound, particularly for nervous tissue. If it accumulates, it leads to coma. The liver, as one of its many jobs, has to get rid of ammonia, and coma is one of the serious symptoms of collapsing liver function as in cirrhosis.

Without a watery wash to get rid of the toxic ammonia, the liver needs to turn it into a soluble, compact and much less toxic compound. In fact it makes **urea**, $CO(NH_2)_2$. The two $-NH_2$ groups come from two ammonia molecules and the rest

---

*Actually we do spend a short early aquatic spell in our careers – *in utero* – and this is exactly how we dispose of excess amino nitrogen in the embryonic phase of life, relying on our mother's metabolism to deal with the problem more finally.

Pain-Free Biochemistry   Paul C. Engel
© 2009 John Wiley & Sons, Ltd

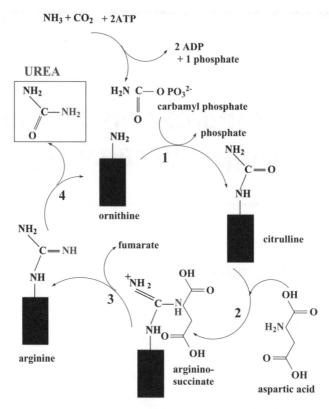

Figure 24.1 The urea cycle. Like the TCA cycle, this pathway was discovered by Hans Krebs, and, as in the TCA cycle, all the intermediates are still there at the end! Their interconversion makes possible the net process, the formation of urea from CO_2 and two amino groups, one fed in as ammonia and the second donated by aspartate.

is donated by carbon dioxide, so that this is a very economical way of disposing of two waste products.

The biochemistry, however, is a little less direct. The pathway to make urea involves a sequence of reactions called the urea cycle. This, like the TCA cycle, was discovered by Hans Krebs. In fact he found the urea cycle first, and it seems likely that one provided the inspiration for the other. In the urea cycle, the molecule of urea is built up on a sort of molecular handle, provided by ornithine. Ornithine is an amino acid, but not one of the 20 that make up proteins. It is not important for you to know its structure, and so in Fig. 24.1 it is just shown as a little rectangular box.

One of the two —NH₂ groups comes in from ammonia, which reacts with CO_2 and ATP to form carbamyl phosphate, $NH_2.COO$-phosphate, one of the feed compounds for the cycle. This then reacts with ornithine, shedding the phosphate and forming another new amino acid, citrulline. The second —NH₂ comes in, not from ammonia but from aspartic acid (see Topic 23). Reaction of aspartic acid with citrulline and another ATP gives argininosuccinic acid (do not worry about either the name or the structure), which promptly breaks down to give arginine, which *is* one of the 20 protein amino acids. The remainder of the aspartic acid skeleton is fed back to the TCA cycle as fumaric acid. In the molecule of arginine (Fig. 24.1), we have in fact assembled the urea structure on our ornithine handle and just need to set it free. The enzyme arginase, which is plentiful in the liver, does this, using a molecule of water to hydrolyse the arginine, and the ornithine is then ready to carry out the sequence all over again.

This set of reactions expends quite a bit of ATP to make sure that the ammonia is efficiently converted into a less toxic product. The liver puts the urea out into the bloodstream, but the process is completed by the kidney, which filters the urea out into the urine for final excretion.

Self-test MCQs on Topic 24

1 Which of the following most correctly describes the role of the urea cycle in humans?

(a) It provides a supply of ammonia needed for synthesis of new amino acids.

(b) It provides a means of safe disposal of excess amino groups from amino acids.

(c) It converts amino acid carbon skeletons into CO_2.

(d) It supplies scarce arginine for protein synthesis.

2 Which of the following describes the involvement of tissues in urea metabolism?

(a) Urea is made exclusively by the liver and excreted by the kidney.

(b) Urea is made throughout the peripheral tissues and broken down in the liver.

(c) Urea is made by the liver and kidney and used by heart and brain.

(d) Urea is made by the kidney and stored in the liver.

3 Ornithine is which one of the following?

(a) A carrier molecule for assembling the components of urea.

(b) The main nitrogenous excretion product in birds.

(c) The donor of amino groups for urea synthesis.

(d) Totally converted into CO_2 and water in the urea cycle.

SECTION 3

Anabolism and Control

Is anabolism just catabolism backwards?

All reactions are reversible. Is this sufficient?

We saw in Chemistry IV that chemical reactions are reversible, so that they can be made to run in either direction, depending on the concentrations of the reactants. Metabolic pathways are not single reactions but strings of linked reactions. Nevertheless, the same principle applies, in theory. So, until about 50 years ago, biochemists tended to assume that in working out the reactions and enzymes responsible for breaking down carbohydrates, fatty acids and proteins they had also worked out the sequence for the reverse processes of biosynthesis. The direction of flow would depend on the ratios of concentrations of starting materials and end products. Realistically, in fact, this would require huge swings in these ratios, whereas living things aim to minimise such swings and keep conditions as constant as possible (homeostasis). Gradually it started to emerge that biosynthesis of fatty acids, say, did not simply reverse the reactions of β-oxidation. Nowadays it seems amazing that biochemists could seriously have thought such an arrangement could work, but hindsight is always a marvellous thing!

Spatial separation

The specific organisation is different for each individual pathway, but nevertheless there are some broad generalisations that can be made. First of all, degradation and synthesis are often physically separate. Fatty acids may be both made and used in the same tissue, e.g. liver, but the two directions of metabolism are found in

Pain-Free Biochemistry   Paul C. Engel
© 2009 John Wiley & Sons, Ltd

different parts of the cell (Topics 20 and 27). This kind of division of labour within the same cell is called **compartmentation**. In other cases, the separation may be even more complete, as with ketone bodies, which are made and used in different organs (Topic 21).

Input of chemical energy

Secondly, even when, as in the conversion of glucose into pyruvate (Topic 13) and *vice versa* (Topic 26),

    1 the processes may take place in the same organ

    2 and the same compartment (e.g. cytosol of liver cells)

    3 and many of the chemical reactions are indeed the same, only in reverse,

there are always some reactions that are different, and this ensures that the change of direction can be achieved without enormous swings in concentration. In particular, the *biosynthetic sequences are always constructed to use more ATP than will be regained in the degradative pathway*. This might seem like bad economy, but it is the price we pay for keeping things efficient.

Isoenzymes

Finally, it turns out that often even the reactions that are common to both the synthetic and the degradative routes have two separate enzyme proteins for the two jobs! In theory, either one should be able to do both jobs; an enzyme has to catalyse both directions of its reaction equally. Nevertheless, there seems to be fine-tuning of some of these enzymes to fit them more ideally for one role or the other. These enzymes, different proteins catalysing exactly the same chemical reaction, are called **isoenzyme**s (Appendix 10).

Self-test MCQ on Topic 25

Which one of the following is an accurate statement about anabolism and catabolism in human cells?

(a) Reversibility of all chemical reactions makes it unnecessary to have separate enzymes for anabolism and catabolism.

(b) Anabolism and catabolism are kept in separate cellular compartments.

(c) Anabolism and catabolism utilise some of the same enzymes but use different reactions for some conversion steps.

(d) Catabolism neatly reclaims all the ATP expended in anabolic reactions.

Making new glucose: gluconeogenesis

We need to be able to make sugars

Gluconeogenesis is just a fancy term for making glucose anew or from scratch, i.e. ending up with sugar, having started with no sugar. There are no 'essential sugars' in the sense that there are essential amino and fatty acids that we must take in through our diet, but this does not mean that we do not need sugars. On the contrary, there are no 'essential sugars' perhaps because sugars are in fact so essential that we have to be able to make them ourselves!

The routes to some of the sugars involved in cell recognition, etc. are beyond our scope, but there are two mainstream routes for sugar synthesis that we have to consider. One, to make pentoses, five-carbon sugars, appears in Topic 28. The other, to make glucose, is our topic for this section. We need glucose as a precursor for some of the other sugars, but above all we need it as a compulsory energy source for various tissues, especially the brain and other nervous tissue. For this reason, our body strives to maintain the blood glucose level at a remarkably constant 80–100 mg/dl, and usually succeeds.

Lactate as a gluconeogenic source

Glucose may be synthesised from products of amino acid breakdown (Topic 23), from glycerol released in the breakdown of fats (Topic 19) or by using products of glycolysis – lactate and pyruvate (Topic 13). This last category might seem a bit futile: breaking down glucose only to build it up again. However, we have to think about this in the context of the whole body. In Topic 13, we looked at the way in

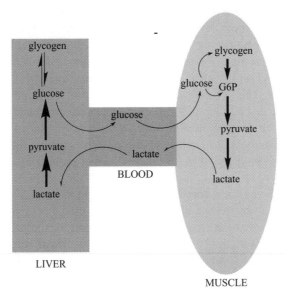

Figure 26.1 Interconversion of glycogen and lactate: opposite directions of flow in skeletal muscle and in liver. The lactate made by muscle during exercise is carried through the bloodstream to the liver, which can reconvert it to glucose and/or glycogen. The glucose can then restock the reserves in muscle. This is known as the Cori cycle.

which skeletal muscle, in an emergency situation, resorts to anaerobic metabolism, converting pyruvate into lactate. What happens to that lactate as it accumulates? It is carried away by the blood supply and so becomes available to other tissues. Some tissues (e.g. heart), in a more aerobic state than the skeletal muscle, may take up the lactate to use as an oxidisable substrate. However, another possibility is that the lactate may find its way to either the liver (Fig. 26.1) or the kidney. Both these tissues have the enzyme machinery to convert lactate back to glucose (and in the case of the liver also onward to glycogen) (Appendix 10 and Box 8).

Box 8 Different enzymes in different tissues

In Topic 26 we met the idea that liver and kidney have the enzyme machinery to convert lactate into glucose, the enzymes of gluconeogenesis. Some of these are just the enzymes of glycolysis used backwards. But some are not. In Topic 21, we met a different relationship between liver and kidney: the liver has the enzymes to make and export ketone bodies; the kidney and also the heart have the enzymes to use them.

This then makes a very important general point. The cells of liver, kidney and heart all have the same starting genetic instructions (see Topics 32–35) and

(Continued)

therefore in theory 'know' how to make all these enzymes, but they do not make them all. Each tissue has gone down its own particular developmental pathway to become the tissue that it is, and that involves, amongst other things, switching on the production of particular enzyme proteins and switching off others, perhaps for ever. This, of course, saves a great deal of waste of energy and chemical resources because each cell only makes the machinery that it needs. The red blood cell does not need to worry about making the enzymes to break down neurotransmitters or to carry out muscle contraction; adipose tissue does not concern itself with enzymes for laying down or mobilising glycogen – and so on.

Apart from giving us a very obvious handle on what each tissue is up to metabolically, this simple fact also offers clinical medicine a powerful tool. If there are enzymes that are characteristic of the heart, or the prostate gland, or particular cell types in the brain, then we should worry if we find these proteins in the bloodstream because these are indicators of cell damage, perhaps as a result of a heart attack or a prostatic or brain tumour. Sure enough, clinical biochemists and chemical pathologists have a battery of tests to exploit these biochemical markers of tissue damage. Another context in which they can be very valuable is organ transplant. The constant danger is that there is tissue rejection through a host immune response. One of the earliest signs will be the appearance of proteins in the bloodstream that are characteristic of the transplanted organ, allowing the clinician to take appropriate action in good time.

The first step is to oxidise lactate back to pyruvate using NAD^+. This is a straight reversal of the LDH reaction (Topic 13). It is also a striking example of the point made in Topic 25 about isoenzymes. Tissues that only need to reduce pyruvate to lactate make almost exclusively type M (**M**uscle) LDH. Those that mainly oxidise lactate make more type H (**H**eart) LDH, and those that do both make a mixture of the two types.

Bypass reactions in the route from pyruvate to glucose

Most of the reactions on the way back to glucose from pyruvate are the reactions of glycolysis, but, as we noted in Topic 25, there need to be some alternative reaction steps to make gluconeogenesis energetically feasible without huge changes in concentrations. There are in fact three 'bypass' steps. Up at the 'top' end, in glycolysis there are two steps that use ATP, converting glucose into G6P and F6P to fructose 1,6-bisphosphate. If we simply reversed those two steps, at least we would get back the ATP that might have been spent earlier. However, we would be struggling to drive the reaction steeply uphill in energy terms, i.e. to drive G6P and ADP back to glucose and ATP would require very high levels of G6P and ADP (and/or very

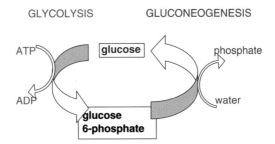

Figure 26.2 Converting glucose 6-phosphate to glucose in gluconeogenesis. The ATP expended in glycolysis is not claimed back, helping the process to work 'uphill' as well as 'downhill'.

low levels of glucose and ATP). By splashing out a bit we can avoid the problem entirely! There is a different enzyme activity, glucose 6-phosphatase, which splits the phosphate off G6P but uses water rather than ADP to receive it (Fig. 26.2). This is now a different overall chemical reaction, and, even though we are still splitting G6P to give glucose, energetically we have turned it from an uphill process to a downhill process. Exactly the same trick is deployed at the step to remake F6P. So these two steps work smoothly in both directions because on the gluconeogenic path we are prepared to forego the ATP that would come with straight reversal of the reactions of glycolysis.

Down at the bottom end, things are slightly different. The conversion of PEP into pyruvate in glycolysis is one of the reactions that *makes* ATP, and so reversing it would split ATP. Superficially that might seem like a good way to make the reaction go, but we have to think why the glycolytic reaction works in the first place. The energy associated with making or splitting PEP is even larger than that associated with ATP, and so reversing the pyruvate reaction in order to make PEP would again be an uphill struggle. Splitting one ATP is not enough here! Nature solves this problem by taking what looks like a pointless detour. It uses two reaction steps instead of one. The first one inserts CO_2 into pyruvate to make oxaloacetate, and the second one chucks it out in a reaction that makes PEP. Why add CO_2 if you do not really want it there? The point of this is that *both* these steps incorporate an energy boost by splitting ATP in the first case and GTP in the second.

So, to summarise, two of out three bypasses work by not making ATP that would be there if one simply reversed glycolysis, and the third does so by using up more ATP/GTP than would be spent by just reversing glycolysis.

Other gluconeogenic precursors

We have just met oxaloacetate as a gluconeogenic intermediate between pyruvate and PEP. But oxaloacetate is also one of the Krebs cycle compounds, and they are all readily interconvertible as you go round the cycle. This means that *any* of the Krebs cycle compounds can be gluconeogenic precursors via oxaloacetate, and anything else (e.g. an amino acid) that feeds into the cycle can also be used to make glucose.

The glycerol from fat breakdown feeds into the glucose breakdown/synthesis pathways at the level of glyceraldehyde 3-phosphate. It is important, however, to note that the fatty acids themselves *cannot* yield glucose. They form acetyl CoA, but in humans there is *no route back from acetyl CoA to pyruvate.* The conversion of pyruvate into acetyl CoA is effectively a one-way valve. Thus, it is easy to convert excess carbohydrate into fat but not the other way round.

Self-test MCQs on Topic 26

1 Which one of the following statements about gluconeogenesis is true?

   (a) Gluconeogenesis in brain allows the tissue to meet its glucose needs from lactate and glycerol.

   (b) Gluconeogenesis in liver is primarily to meet the energy needs of other tissues.

   (c) The heart is a major provider of lactate as a precursor for gluconeogenesis.

   (d) Adipose tissue supports gluconeogenesis by providing fatty acids and hence C2 units as acetyl CoA.

2 Which of the following is not a gluconeogenic precursor?

   (a) acetoacetate

   (b) oxaloacetate

   (c) pyruvate

   (d) glycerol.

TOPIC 27

Fatty acid biosynthesis

Similarities and differences between oxidation and synthesis of fatty acids

We saw in Topic 20 how long-chain fatty acids are broken down into two-carbon chunks by a process of oxidation, hydration, oxidation and splitting with CoA. The dwindling chains are chemically activated for this process by attachment to the S atom of CoA. If our tissues need to make new fatty acids instead of oxidising them as an energy source, to what extent can the process be reversed? In broad outline there are similarities. In fatty acid synthesis, acetyl CoA units provide the raw material, the growing chain is activated by attachment to a sulphur atom, and there are repeated cycles of condensation, reduction, dehydration and reduction. At first sight, this seems like a simple reversal of the steps of β-oxidation. However, in line with our generalisations in Topic 25, there are multiple differences as follows:

1 Beta-oxidation occurs mainly in mitochondria, whereas fatty acid biosynthesis is carried out in the cytosol.

2 The activating sulphur attachment is not to CoA but to a much larger object, a small protein, called the acyl carrier protein (ACP). In both cases, CoA and ACP, the sulphur-containing bit is derived from dietary pantothenic acid (a B-vitamin).

3 In the oxidative pathway the two oxidations are carried out, one by FAD and one by NAD^+. In the biosynthetic pathway, however, both reductions are carried out by NADPH (see Topics 10 and 28). In the case of replacing $FADH_2$, this is an energetic consideration, in that NADPH is a stronger reducing agent (the equilibrium point will lie further over towards reduction). In the case of replacing

Pain-Free Biochemistry   Paul C. Engel
© 2009 John Wiley & Sons, Ltd

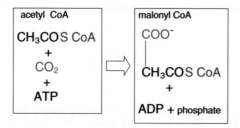

Figure 27.1   Formation of malonyl CoA.

NADH, this is an example of a broad metabolic separation between the roles of two chemically very similar coenzymes: chemically speaking it could just as easily have been the other way round, but Nature's decision is that NAD^+ is used throughout energy-yielding catabolism and NADPH is used throughout biosynthesis (but see also Topic 28).

4   There is an additional energetic drive provided in a way that is very similar to what we have just seen in gluconeogenesis in the conversion of pyruvate into PEP. Although an acetyl CoA unit is used every time another two carbons are added to the growing chain, each of them is first converted into malonyl CoA. This involves the splitting of ATP and adds an extra —COOH, derived from CO_2 (Fig. 27.1). The enzyme for this step uses a cofactor we have not met previously, **biotin**. At the next step, the condensation reaction, the CO_2 is promptly thrown out again, just as with PEP. The initial ATP-driven reaction upgrades the feed-in unit for the condensation so that the reaction can run downhill energetically. The CO_2 is not needed as part of the growing structure; it is an excuse for building in the input of chemical energy from the splitting of ATP.

5   Finally, several of the separate enzyme catalysts in this pathway are physically joined to one another as giant proteins, so that the whole assembly is like an efficient production line, with molecules being handed on for the next step.

Other starting points

Fat synthesis is not totally tied to starting from scratch with two-carbon units. First of all it can incorporate readymade fatty acids into triglyceride, so that the fatty acid composition of our own fats will be influenced by the composition of the fats we eat. Also we can elongate a readymade fatty acid by adding further C2 units onto it. We do this with some of the essential fatty acids in our diet. Down at the end furthest from the —COOH they have patterns of double bond that we cannot make ourselves. These patterns can remain intact as we convert, say, a C20 fatty acid into C22 or C24.

Self-test MCQs on Topic 27

1 Which of the following is not used in fatty acid biosynthesis?

(a) CO_2

(b) CoA

(c) Carnitine

(d) Acyl carrier protein

2 Which of the following correctly describes the role of biotin in fatty acid biosynthesis?

(a) It carries the growing acyl chain.

(b) It supplies the reducing equivalents.

(c) It ferries finished fatty acids across the cell membrane.

(d) It assists in formation of malonyl CoA.

TOPIC 28

Providing reducing power: NADPH and the pentose phosphate pathway

How do we make NADPH?

The fatty acid biosynthesis pathway presents us with the first example of an interesting puzzle. If all the oxidative metabolism of foodstuffs is tied into making NADH (to be reoxidised and to make ATP), how do we get a supply of NADPH for biosynthesis? What reduces the $NADP^+$?

Conversion of hexose into pentose

The answer is that there are a few 'specialist' dehydrogenases that go against the usual rule, in that they are catabolic enzymes but use $NADP^+$. The most important are a pair associated with what is called the 'pentose phosphate pathway' (PPP). This pathway can do various things depending on what is fed into it and what is required, but under typical conditions in humans it offers an alternative route for metabolising glucose. In cells that have the PPP enzymes, G6P becomes a branch point because it can either go down through glycolysis via F6P or it can take the alternative route. G6P dehydrogenase uses $NADP^+$ (and will not accept NAD^+) to oxidise the end aldehyde group of G6P to an acid —COOH group, making 6-phosphogluconate. This then becomes substrate for a second dehydrogenase, which also uses $NADP^+$, and in carrying out its oxidation also removes one of the six carbons as CO_2, making the 5-carbon ketose sugar ribulose 5-phosphate (hence

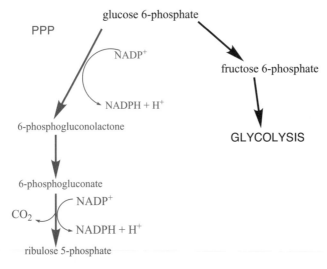

glucose 6-phosphate

PPP

NADP$^+$

fructose 6-phosphate

NADPH + H$^+$

6-phosphogluconolactone

GLYCOLYSIS

6-phosphogluconate

NADP$^+$

CO_2

NADPH + H$^+$

ribulose 5-phosphate

Figure 28.1   NADPH-forming steps of the pentose phosphate pathway.

PPP). Each glucose molecule fed into this pathway thus gives rise to two molecules of NADPH to be used for biosynthesis (Fig. 28.1).

Recycling pentose

What follows is a kind of chemical juggling act. First of all there are two enzymes that rearrange the structure of ribulose 5-phosphate, either converting it into the corresponding five-carbon aldose sugar ribose 5-phosphate or else swapping the —H and —OH on carbon 3 to make the alternative ketose xylulose 5-phosphate. This then creates an interchangeable pool of three different pentose phosphates to take part in the horse-trading catalysed by two further enzymes, transketolase and transaldolase. Both these enzymes transfer chunks between sugar molecules. Transketolase shifts a two-carbon unit, carbon 1 and carbon 2 from a ketose sugar, such as xylulose 5-phosphate, to a receiving aldose sugar such as ribose 5-phosphate. At this point, biochemistry becomes arithmetic!

$$5 + 5 = 3 + 7$$

In other words, all that is left of xylulose 5-phosphate is the three-carbon stump, which is none other than our old friend glyceraldehyde 3-phosphate. At the same time, we have made a seven-carbon ketose sugar. Next we deploy transaldolase, which transfers three-carbon units, and use it on the same 3 + 7 pair. Taking carbons 1–3 from the seven-carbon sugar and transferring them to the three-carbon sugar, we discover that:

$$3 + 7 \text{ also} = 6 + 4$$

The six-carbon product is F6P. This is back to mainline glycolysis, and we can put the F6P on one side while we go on to deal further with the four-carbon sugar. At this point we dip into our pentose phosphate pool a third time to fish out one more molecule of xylulose phosphate and we return to transketolase for another two-carbon transfer.

$$4 + 5 = 6 + 3$$

This makes a second molecule of F6P plus a molecule of glyceraldehyde 3-phosphate.

Like all juggling acts, this leaves one a bit dizzy, but if we look back carefully at what has happened it is really very crafty! Overall we have taken three molecules of hexose phosphate. Initially, we have converted these into three molecules of pentose phosphate plus three molecules of CO_2. In the process, we have achieved our initial objective by reducing six molecules of $NADP^+$ to NADPH. Then we have used two enzymes to reshuffle the atoms of the three pentose molecules to give us back, in effect, $2\frac{1}{2}$ hexose phosphate molecules, back on the mainline glycolytic pathway (as we have seen in Topic 26, triose phosphates are readily reconvertible to hexose phosphates).

PPP to supply pentose

This is a pathway that is almost infinitely flexible in its possibilities. Clearly, since some of the carbon atoms are back where they started at the level of C6 sugars, one could simply go back round again, each time topping up with some fresh hexose phosphate. On the other hand, we could feed some of the proceeds on down through glycolysis to make some ATP. A third very important possibility, however, is that we could milk the pathway for another product. So far we have only considered it as a way of making NADPH reducing power available for biosynthesis. However, pentose phosphates are not just boring metabolic intermediates; they are absolutely vital building blocks. We have already met a whole series of key compounds in metabolism that incorporate pentose in their structure – ATP/ADP, NAD^+/NADH, $NADP^+$/NADPH, FAD/$FADH_2$ and CoA. We might, of course get some of the pentose we need through our diet, but, if not, we need to be able to make it. However, there is another critically important part of our life chemistry that is built round pentose. In Topics 32–34, we shall meet DNA and RNA, all-important molecules built round pentose units. This makes the PPP absolutely essential for growth and cell division.

Distribution of PPP activity

The PPP and its enzymes are not to be found in all tissues but are prominent in those tissues that are active in biosynthesis. Liver, for example, may feed up to 30% of its glucose through this path. Likewise, adrenal cortex, constantly involved in biosynthesis of steroids, also needs NADPH. Muscle, by contrast, is not a biosynthetic tissue, and therefore has no need of the PPP enzymes. Some tissues carry out active biosynthesis only intermittently. A striking example is mammary gland: as a mother's body prepares to feed her offspring the enzymes of the PPP are switched on in the breast tissue. After weaning they gradually disappear.

Self-test MCQs on Topic 28

1 Which one of the following statements about the PPP is untrue?

  (a) This pathway makes NADPH for biosynthesis.

  (b) This pathway makes ATP for biosynthesis.

  (c) This pathway is active in the liver.

  (d) This pathway supplies pentose for nucleotide synthesis, e.g. for DNA and RNA.

2 Which of the following describes the action of transketolase in the PPP?

  (a) It converts a six-carbon sugar into a five-carbon sugar plus CO_2.

  (b) It interconverts different five-carbon sugars.

  (c) It catalyses NADPH production by oxidation of ribulose 5-phosphate.

  (d) It moves two-carbon units between sugar phosphates.

CHEMISTRY X

Isotopes

A reminder of atomic structure

Way back in Chemistry II we looked at the way in which each of the chemical elements, hydrogen, oxygen and so on, is defined by its atomic number, which is the number of protons it has in its nucleus (and also the number of electrons circulating round that nucleus, unless and until the atom starts interacting with other atoms, sharing, receiving or donating electrons to form molecules). We also saw that the nucleus contains a number of electrically neutral particles, neutrons, the number usually being equal or roughly equal to the number of protons. Table II.1 in Chemistry II shows that the hydrogen atom has one proton, one electron and no neutrons (in spite of the statement just made), and that carbon has six each of protons, electrons and neutrons.

The number of neutrons can vary

In fact, although the description above is the truth, it is not quite the whole truth. Most elements have a 'favourite' atomic structure, but it may not be the only model available! If you have a slightly different number of neutrons, it does not alter the atomic number (which is determined by the number of protons) and it does not alter the number of electrons that the atom presents to the outside world and therefore the chemistry it will take part in. Even hydrogen, although the vast majority of atoms are just as we have described them, includes a small proportion of weightier individuals that do have a neutron in the nucleus, and a still smaller proportion that manage to squeeze in two neutrons! These alternative forms of the same element are called **isotopes**. In the case of hydrogen (only), the additional forms have been given special names *and* symbols: hydrogen with one neutron is called deuterium (D) and

Pain-Free Biochemistry   Paul C. Engel
© 2009 John Wiley & Sons, Ltd

hydrogen with two neutrons is called tritium (T). All isotopes of any element can be represented with a superscript number (^{1}H, ^{2}H and ^{3}H for the three isotopic forms of hydrogen), which indicates the relative atomic mass of that particular atomic form, e.g. tritium has one proton and two neutrons, adding up to a relative atomic mass of 3.

Unstable isotopes: radioactivity

In some cases these alternative forms of an element are stable and unchanging, but some isotopes have nuclei that are not very stable and therefore tend to disintegrate and convert into more stable products. These are said to be radioactive because invariably the disintegration is accompanied by emission of some form of radiation. We have various sensitive measuring devices that can detect this radiation.

For a couple of centuries, chemists regarded atoms and the chemical elements as the unbreakable minimum building blocks; they could be recombined in different partnerships (chemical compounds) but never changed or destroyed. Radioactive decay, however, involves changes in an atomic nucleus that alter one element into another! Thus, if we take the unstable atom of tritium as an example, we start with one proton and two neutrons, but in the radioactive decay event one of the neutrons is converted into a proton plus an electron. This conserves the mass of 3 and also the net electric charge, but now we have two protons and one neutron in the nucleus. This means we have changed atomic number and so we no longer have a hydrogen atom. Instead we have a helium atom. Note, however, that this helium isotope is not the common version of helium either. The predominant helium isotope has two protons and two neutrons and hence a relative atomic mass of 4.

Although we have said that tritium is unstable, the chances of any one particular tritium atom disintegrating over a period of, say, 1 h would be very low. Relative stabilities of various isotopes are expressed in terms of their half-lives. This refers to the time it would take for exactly 50% of the material to undergo radioactive decay. In the case of tritium the figure is approximately 12 years.

Carbon 14

Another very important isotope for biochemistry has a much longer **half-life** than that. The stable majority isotope of carbon, ^{12}C, has six protons and six neutrons, but ^{14}C, with eight neutrons, is radioactive. Just like tritium, it can convert one of the extra neutrons into a proton and an electron, turning it into a nitrogen atom with seven protons and seven neutrons. This occurs so slowly that the half-life of ^{14}C is about 5700 years! This has made it important for dating in archaeology and palaeontology.

Artificial enrichment of particular isotopes: metabolic labelling

Why are these isotopes important in biochemistry and medicine? The isotopes we have mentioned occur at very low 'natural abundance', e.g. in the world around us only about 1 carbon atom in 10^{12} (a million million) is ^{14}C. However, with the advent of nuclear physics and specifically the Manhattan Project, the atomic bomb project in World War II, radioactive isotopes started to be produced artificially, and this meant that chemical compounds could be radioactively 'labelled', either uniformly (e.g. ^{14}C in every carbon position) or selectively (i.e. with radioactive enrichment in particular positions). In the case of carbohydrate metabolism, it was possible to study the relative importance of glycolysis and PPP by comparing the release of radioactivity from glucose, specifically labelled either in carbon 1 or in carbon 6. If you look at Topic 28, you will see that in the initial reactions of the PPP the CO_2 that is produced comes entirely from the C1 position. Over time, as the later molecular rearrangements come into play, C6 atoms could also eventually be released but not initially. On the other hand, if you revisit Topics 13 and 14, you will see that, because the sugar phosphate is split down the middle into two triose phosphate halves that are then handled identically, CO_2 released in the oxidation of pyruvate to acetyl CoA will be derived equally from C1 and C6. This allows biochemists to assess the relative activities of PPP and glycolysis in different tissues or in the same tissue over time. This is how it was possible to estimate (Topic 28) that 30% of glucose breakdown in liver is via PPP.

Appendix 9 describes how Knoop followed metabolism of fatty acids (Topic 20) by attaching a bulky and recognisable chemical label, the phenyl group. This was ingenious, successful but really a rather risky strategy because it depended on the relevant enzymes 'not noticing' the bulky label. From 1945 onwards a biochemist would automatically have looked instead to isotope labelling as the way to do that sort of experiment.

Use of stable isotopes

In more recent years, it has become less essential to use radioactivity, which, though very useful, is also hazardous. As we have mentioned, there are also stable, rare isotopes. In the case of hydrogen we have deuterium; in the case of carbon we have carbon 13 (seven neutrons). It is also possible to label a biochemical compound with a stable isotope. Some of these isotopes (^{13}C being one) can be monitored by nuclear magnetic resonance (NMR); all of them can be detected by mass spectrometry, which, as the name implies, relies on measuring the differences in mass of different molecules or fragments of molecules. The improved sensitivity of these two techniques in recent years has enhanced the usefulness of stable isotopes.

Isotopes thus give us a very powerful method for tracking the metabolic fates of compounds in our bodies.

Self-test MCQs on Chemistry X

1 Isotopes of an element are due to which of the following?

(a) Varying numbers of protons

(b) Varying numbers of electrons

(c) Varying numbers of neutrons

(d) Varying numbers of all three types of particle.

2 Isotopes of an element are which of the following?

(a) Always radioactive;

(b) Sometimes radioactive;

(c) Never radioactive.

3 Isotopes are valuable to us because

(a) they stimulate the immune system

(b) they are needed in trace amounts for enzyme action

(c) they allow us to label and follow compounds in the body

(d) they provide cells with an energy source.

TOPIC 29

Red cells and white cells: defence against reactive oxygen and reactive oxygen as defence!

Oxygen as a threat

Thus far we have considered NADPH and the PPP only for growth and biosynthesis (Topic 28). NADPH, however, has a significant and different role even in certain cells that have forgotten how to divide and certainly are no longer able to carry out biosynthesis. An outstanding example is the red cell. Mature erythrocytes in humans have no cell nucleus and so cannot carry out any of the functions associated with a nucleus. Once they are launched on their short careers, with their full complement of enzymes and of haemoglobin, there is no more synthesis of new proteins in these cells. However, the cells and their proteins have to survive for several months of highly dangerous work. The danger comes from their cargo, oxygen. We think of oxygen as our indispensable lifeline, but oxygen is a very chemically reactive molecule. In particular, it is capable of being activated within the cell to several desperately dangerous forms – superoxide, hydrogen peroxide and hydroxyl radicals. These are so reactive that they are capable of inflicting fatal damage on membranes, on DNA, on proteins, etc.

Pain-Free Biochemistry   Paul C. Engel
© 2009 John Wiley & Sons, Ltd

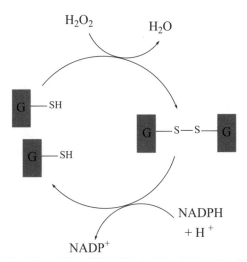

Figure 29.1 The role of glutathione. The active form of glutathione is the reduced, —SH form. The spent, oxidised —S—S— form has to be recycled, being reduced by NADPH.

Glutathione

It is vital to have a ready defence against reactive oxygen species, like fire extinguishers on an oil tanker. One of several elements in the defence repertoire is a compound called **glutathione**. It is in fact a tripeptide (see Topic 4), but the important thing is that one of the three amino acids that make up this tripeptide is cysteine, which has an oxidisable —SH group. Glutathione acts as a scavenger for hydrogen peroxide (H_2O_2) with the help of the enzyme glutathione peroxidase. If we represent glutathione for convenience as **G—SH**, the reaction converts H_2O_2 into water and, in doing so, oxidises 2 **G—SH** molecules, linking them to form **G—S—S—G**. The glutathiones sacrifice themselves to render the peroxide harmless. This is where NADPH comes in. Glutathione would be of limited use if it ran out of steam after one use. Far better if the kamikaze molecules could rise from the dead to fight another day. In fact, there is a system to replenish the defence by keeping the glutathione constantly reduced. Another enzyme, glutathione reductase, has the specific job of reconverting **G—S—S—G** to its 2 **G—SH** halves by reducing it with NADPH (Fig. 29.1).

PPP in red cells and G6P dehydrogenase deficiency

Because of this requirement for NADPH to reduce glutathione in red cells, they contain the enzymes of the PPP. We can also see what happens when this system fails because a genetic deficiency of the first PPP enzyme, G6P dehydrogenase, is the most common of all known inherited diseases in humans, affecting about 400 million people. It tends not to be fatal (otherwise it could not be so widespread) but nevertheless causes haemolytic anaemia, often triggered by drugs. One of the reasons we can usually survive this condition is that the nature of the problem tends to be that the mutant enzyme is not rugged enough to stand up to months of service in the red blood cell. In other cells, such as those of the liver, if the enzyme molecules wear out they can be replaced, but not in the red cell with no nucleus. So this condition is bad news for the red blood cell, but it is highly prevalent, and non-fatal though debilitating, precisely because it *only* affects the red cell.

This still does not quite explain why this genetic disease is so common. In fact, in tropical countries human red cells often carry unwanted passengers in the form of the malaria parasite, *Plasmodium*. These hitch-hikers are even more upset by reactive oxygen species than the red cells themselves, so that the genetic disease actually has a positive side effect of conferring resistance against malaria.

Oxidative damage following ischaemia

Another situation where oxidative damage is clinically important occurs during recovery from an ischaemic episode. Ischaemia is the term that describes the interruption of normal blood supply to a tissue, as happens, for example, during a heart attack or a stroke. Without a blood supply the tissue is deprived of an adequate oxygen supply. This undermines all the energy-requiring functions of cells, which therefore become progressively damaged. This makes sense, but where does oxidative damage come in without oxygen? The problem arises after medical intervention, e.g. to dissolve a blood clot or dilate a coronary artery: reperfusion of the damaged tissue, restoring the oxygen supply, tends to form reactive oxygen species that accentuate the tissue damage. Now that this is understood, these effects can be minimised with the help of antioxidants such as glutathione.

Reactive oxygen species as a defence: respiratory burst in white cells

We have just seen NADPH as a defence against reactive oxygen species. Now for what may seem the ultimate contradiction, and again in our blood cells! What about NADPH to *generate* reactive oxygen species? First of all why would any responsible cell want to do that? In a sense, we had a hint above with the idea that reactive oxygen can kill malaria parasites. That was more by accident than design because red cells are not normally set up to repel invaders in that way. However, the blood contains other cells that are there for precisely that purpose. Perhaps not so much to repel them as to swallow them whole, kill them and digest them! Among the white blood cells are phagocytes, so named because they gobble up unwelcome visitors ('phago' comes from the Greek word for 'eat'). These cells have a specialised respiratory chain. In Topic 15, we saw how the mitochondrial respiratory chain uses oxygen to reoxidise NADH in a stepwise sequence that generates ATP. The phagocytes, like other aerobic cells, can do that too, but in addition they have their own separate system that reoxidises NAD*P*H at the expense of oxygen. This machinery makes no ATP, and instead of producing harmless water as the end product of the oxidation it 'deliberately' produces reactive superoxide, which in turn is converted into hydroxyl radicals. Rather than a low-level leakage, this is a sudden and massive production of a toxic agent. This process is known as the **respiratory burst** and is designed to kill engulfed bacteria and other invaders.

Self-test MCQs on Topic 29

1 Which one of the following statements about glutathione is false?

(a) Glutathione is a sugar produced by the PPP.

(b) Glutathione can readily be oxidised.

(c) Glutathione has an —SH group essential for its function.

(d) NADPH is important in maintaining the status of glutathione in the cell.

2 Which one of the following statements about the respiratory burst is true?

(a) It refers to lung damage following hyperventilation.

(b) It refers to the activation of the mitochondrial electron transport chain during exercise.

(c) It refers to rupture of the oxygen-carrying red cells after membrane damage.

(d) It refers to specialised respiratory activity in neutrophils to produce reactive oxygen species.

3 G6P dehydrogenase deficiency is credited with which one of the following?

(a) Eradication of smallpox.

(b) Resistance to malaria.

(c) Tolerance of HIV infection.

(d) Immunity against yellow fever.

TOPIC 30

The need for metabolic control

The problem

We have seen in earlier chapters that our bodies have metabolic pathways both for breaking down substances like sugars and fatty acids and for making them from smaller molecules. These pathways in some cases overlap, in the sense that they may use some of the same reactions in reverse. The direction of the net flow at each of these steps will depend simply on the concentrations of the reacting substances (see Chemistry IV). However, there are also reactions that are unique either to the catabolic or to the biosynthetic sequence. In the case of conversion of glucose into pyruvate, or pyruvate into glucose, we saw that three steps in glycolysis have a bypass reaction in gluconeogenesis. This creates a potential problem but also offers a solution.

One obvious point to make first of all is that it would be stupid for the same cell or tissue to be both making glucose or fat and, at the same time, breaking it down. One way of avoiding this pointless situation is to divide up tasks between tissues. Thus, skeletal muscle, for example, does not need to confront the problem of reconciling glycolysis and gluconeogenesis: it simply does not carry out gluconeogenesis and does not make the special enzymes associated with gluconeogenesis. However, the tissues that do carry out gluconeogenesis, liver and kidney, also have the glycolytic enzymes. So what happens here?

The situation looks potentially disastrous if we close in on an individual step, e.g. the interconversion of glucose and G6P. We have two opposing reactions, each rigged to work satisfactorily in energy terms. So converting glucose into G6P (the glycolytic reaction) works because the reaction also converts ATP into ADP,

Pain-Free Biochemistry   Paul C. Engel
© 2009 John Wiley & Sons, Ltd

providing an energy 'push' to make the overall process favourable. Coming back the other way, the gluconeogenic reaction works precisely because we *do not* involve ADP and ATP, but just split off the phosphate with water. In fact, if we put both reactions together in the same cell compartment, the levels of glucose and G6P might stay the same, going round and round the loop, but ATP would be continuously converted into ADP and phosphate, squandering the cell's carefully hoarded energy store! This is rather like sitting in a car with one foot on the accelerator and then putting the other foot down on the clutch: suddenly the engine does not need to do any work, but it keeps racing, burning fuel as fast as it can. Like the burning of petrol or diesel, the splitting of ATP is an energetically favourable process that will race away given half a chance.

In the case of fatty acid metabolism, any problems of this sort are avoided because, even though both fatty acid oxidation and fatty acid synthesis can occur in the same cell (e.g. in liver cells), they are kept safely apart in different cellular compartments. With glucose metabolism, however, the enzymes for both glycolysis and gluconeogenesis are in the cytosol! So do we burn up all our ATP?

Clearly, this is not a situation that the cell could tolerate. Cycles such as the one just described for glucose and G6P are called **futile cycles** and they have to be prevented or controlled. The realisation of this fact led to the recognition of 'metabolic regulation', a process that involves coordinated up- and down-regulation of opposing enzyme reactions. This can happen in more than one way, as we shall see in Topic 42, but for the moment the idea to grasp is that enzymes do not automatically work 'flat out'. At key regulatory points in metabolism there is in effect a 'volume control'. During the 1960s it was realised, initially through studying the control of glycogen synthesis and breakdown and the effects of the hormone adrenaline, that this control of enzymes is often brought about by a chemical modification of the enzyme protein itself. Specifically, this is brought about by **phosphorylation** or 'dephosphorylation', which means simply adding a phosphate group onto an amino acid side-chain or removing it. The phosphate comes, as usual, from ATP or GTP. Enzymes that transfer phosphate from ATP or GTP are called **kinases** and so the enzymes in charge of this traffic regulation are collectively known as **protein kinases** (because the substance they are converting is an enzyme protein). Conversely, there are also protein **phosphatases** that rapidly remove the phosphate. It is not important for you to learn the detail of this, but it is important to understand that these so-called 'regulatory enzymes' in effect have an on/off switch so that they are catalytically active in one state and inactive in the other. In order to avoid a futile cycle, all the cell has to do is to ensure that when the glycolytic enzyme is 'on' the gluconeogenic enzyme is 'off' and *vice versa*. This is exactly what happens (see also Topic 42).

Once this general principle was established, it was gradually realised that this kind of 'switch' is used over and over again in a wide variety of processes in the body, not only all sorts of metabolic pathways but also such processes as cell division, signalling of various kinds, etc.

Self-test MCQ on Topic 30

Which one of the following statements is true?

(a) Anabolic and catabolic enzymes are never found in the same cellular compartment.

(b) Cycling metabolites through anabolic followed by catabolic sequences enhances ATP production.

(c) Anabolic and catabolic enzymes responsible for interconverting a pair of metabolites are subject to opposite on/off control.

(d) The cell avoids futile cycles by ensuring that the same enzyme reaction is used in catabolic and anabolic steps.

TOPIC 31

Relationship of fats and carbohydrates: use by different tissues

Good and bad foods

Not only in a professional context but also in our everyday private lives, we are bombarded with advice and propaganda about individual foods and about diet regimes. With packets of sugar boasting of being 'zero fat' and tubs of lard 'zero carbs', we could be forgiven for thinking that starvation is the only healthy state, although proteins have not yet been put on the hit list! Both for us and for patient care, it is important to have some notion of the importance of the major food sources for different tissues of our bodies and the way they relate to one another.

It is worth bearing in mind that in human history the high carbohydrate diet can only have become widespread since farming began, less than 10 000 years ago, and certainly some hunting communities have lived healthily until very recently on a diet high in protein and fat and very low in carbohydrate.

In trying to work out what we really do or do not need, the first major point to appreciate is that, although initially we try to get a grasp of the typical metabolism of a typical cell-type or tissue, *there is no such thing as a typical human cell*. Each tissue has its own peculiarities and specialisations, and when it comes to feeding, some tissues are very versatile and easy to please while others are exceedingly picky. Most, if not all, human cells can use glucose, but some tissues have an absolute requirement for it. Most important in this context is the brain, and nerve cells in general. Because of this absolute requirement, if the blood glucose level drops much below about 3 mmol/l a person will go into coma, and a key feature of our metabolic regulation is the attempt to maintain a constant blood sugar level. It is known that in

Pain-Free Biochemistry   Paul C. Engel
© 2009 John Wiley & Sons, Ltd

conditions of long-term starvation the brain is capable of some adaptation, enabling it to use ketone bodies, but in most normal circumstances this does not arise.

There are other tissues that only use glucose, and two interesting cases, already touched on in Topic 13, are the red cell and the lens of the eye. Both are living tissues, although not very active metabolically. Neither has heavy energy demands, and both make do with the frugal supply from anaerobic glycolysis, i.e. they convert glucose into lactate and do not use oxygen. Neither tissue has mitochondria. The red cell, of course, has a primary function of supplying oxygen to the rest of the cells of the body, and it makes good sense that it should not raid the supply on the way. The function of the lens depends critically on being transparent, and neither blood vessels to supply oxygen nor mitochondria to permit its use would help transparency. What is very striking in the case of the eye is that within a few millimetres one finds the retina, which, by contrast, is one of the most energetically active tissues of the body and depends on a good blood supply to maintain its aerobic metabolism!

The upshot of this is that clearly we must have a source of glucose. Do we need to obtain it from carbohydrate foods? Not in fact, since, as we have seen in Topic 26, the kidney and liver can make glucose from a variety of precursors, including various amino acids derived from protein breakdown and glycerol from fat breakdown. Once made, glucose can also be converted to glycogen for storage in liver and muscle. On the other hand, as we saw earlier, there are requirements for both essential fatty acids and essential amino acids, and these have to be supplied either as such or, more likely, from fat and protein.

Self-test MCQs on Topic 31

1 Tissues that have to meet their energy requirement through glycolysis include which of the following?

   (a) The erythrocyte and the brain.

   (b) The kidney and the brain.

   (c) The erythrocyte and the liver.

   (d) The kidney and the liver.

2 If blood glucose drops to 2 millimoles per litre, which of the following will happen?

   (a) The person will go blind.

   (b) The person will lose consciousness.

   (c) The person's red cells will burst.

   (d) The person will lose weight.

SECTION 4

Genes and Protein Synthesis

TOPIC 32

The idea of genes

Instructions to make enzymes

As we have seen, metabolism and all the various other life processes are carried out by the action of enzymes, each one having its own unique amino acid sequence. Each type of cell will have many different enzymes, but on the other hand many enzymes present in one type of cell will be missing in another: red blood cells, for example, do not have the enzymes of the Krebs cycle – they have no mitochondria to house them! Even in a single tissue, different enzymes may be found at different times and in different physiological states: as we mentioned in Topic 28, the PPP enzymes are only produced when required in biosynthetic tissues. So somewhere there has to be a set of instructions for making all these enzymes, and it must also be possible either to use the instructions or to ignore them as required. These instructions are called **genes**.

Inheritance of similarities and differences

We have known about genes as a precise concept since the 1850s when Mendel discovered that his pea plants could pass on characteristics from generation to generation in a mathematically exact way. It is also genes that make cats have kittens rather than ducklings or puppies. So genes are passing on the sameness between generations. At the same time some cats are black, others white and others again tabby, and genes are responsible for these differences in cats, just as for the differences between Mendel's peas. So what they are handing on apparently is a very close but not perfect identity. Otherwise everyone we passed in the street would be like our identical twin!

Pain-Free Biochemistry   Paul C. Engel
© 2009 John Wiley & Sons, Ltd

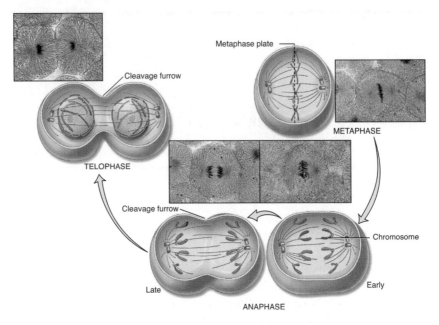

Figure 32.1   Cell division. Under the microscope one of most striking features of cell division is the nuclear division that precedes it. Reproduced from Tortora and Derrickson (2009) *Principles of Anatomy and Physiology*, 12th edn, Wiley International Student Version, New York. © 2009 John Wiley & Sons. Reprinted with permission of John Wiley & Sons, Inc.

Early in the twentieth century, Sir Archibald Garrod applied the same sort of thinking to human disease. He studied various inherited diseases where affected individuals fail to metabolise certain amino acids correctly. Such diseases can result in serious mental impairment but can be recognised through the appearance of unusual metabolites in the urine. Garrod thought about this and came to the conclusion that genes must carry the instructions for individual enzymes. He was absolutely right, and what is remarkable about his deduction is that he made it at a date when science was ignorant of the chemical nature of both enzymes and genes!

Where and what are genes?

One thing seemed reasonably clear, namely that the instructions must be contained in the cell nucleus. Watching cells getting ready to divide leads one irresistibly to this conclusion (Fig. 32.1) Down the microscope one can see worm-like structures, the **chromosomes**, gathering in the middle of the cell, dividing in two and then separating off to form two new nuclei, before the wall comes across to divide the two new daughter cells. So this made it seem highly likely that the chromosomes contained the genes. But what were they chemically? For a long time it seemed clear that the magic substance must be protein because the only other substance around in substantial amounts in the cell nucleus was a very monotonous material called **nucleic acid**. Proteins' 20 amino acids are relatively close to the 26 letters of our alphabet, which we know can carry large amounts of complex information. Unlike proteins, the main nucleic acid in the nucleus appeared to be made of just four different kinds of building block.

In today's computer age, we might be less inclined to dismiss the possibility of a useful coding language with four letters. After all, our computers work on binary code, i.e. a two-letter language, Yes–No and On–Off. In the 1930s and 1940s, most biochemists, however, regarded nucleic acids as likely to be just packing or structural material.

Self-test MCQs on Topic 32

1 Which statement below is true?

(a) Enzymes contain the information to make genes.

(b) Genes contain the information to make chromosomes.

(c) Genes contain the information to make enzymes.

(d) Enzymes contain the information to make chromosomes.

2 A person's genes are inherited from which of the following?

(a) One parent or the other but not both.

(b) The mother only.

(c) Both parents.

(d) The parent of the same gender.

TOPIC 33

The chemistry of genes: DNA and the double helix

Chemical make-up of nucleic acids

In due course biochemists came to realise that nucleic acids, seemingly monotonous in their composition, are not just packing material! First of all, what exactly is this monotonous structure? The major nucleic acid of the cell nucleus is DNA. DNA is found as very long molecules made up of chains of nucleotide (see Appendix 5) units that are linked, sugar to phosphate to sugar to phosphate and so on. Off each sugar there dangles also a 'nitrogenous base', i.e. a single or double chemical ring structure (Fig. 33.1), **adenine**, **guanine**, **cytosine** or **thymine**, very often abbreviated as A, G, C and T. The basic chemical structure of these molecules is rather like a charm bracelet, with the sugar phosphate backbone as the chain and the nitrogenous bases as the charms.

Chargaff's rule

One curious feature of DNA was noted by Erwin Chargaff, who carefully studied the chemical composition of nucleic acids from many different biological species. There was a surprisingly wide variation in the actual percentages of A, G, C and T between different species, but, oddly enough, the percentage of G was always very close to that of C, and that of A to that of T. Chargaff's rule offered a massive clue, but no one apparently spotted what it meant at the time.

Pain-Free Biochemistry   Paul C. Engel
© 2009 John Wiley & Sons, Ltd

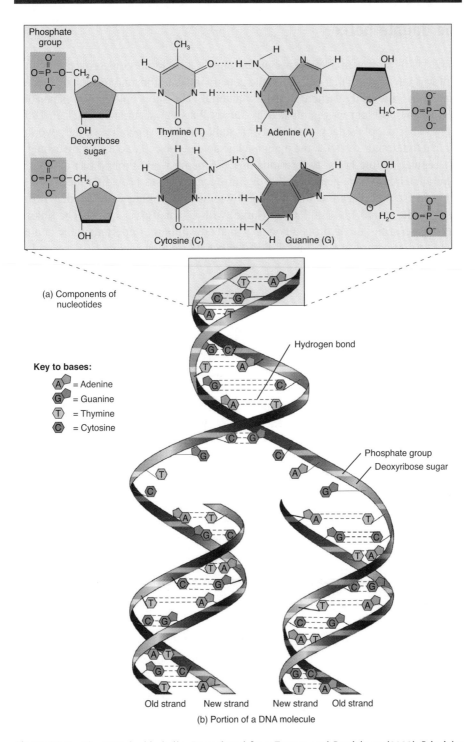

Figure 33.1 The DNA double helix. Reproduced from Tortora and Derrickson (2009) *Principles of Anatomy and Physiology*, 12th edn, Wiley International Student Version, New York. © 2009 John Wiley & Sons. Reprinted with permission of John Wiley & Sons, Inc.

The double helix

In the 1940s, however, a key experiment done on a bacterial virus proved that after all it was nucleic acid that contained the genetic information, and this led on to keen interest in solving the structure of this material. The short but historic paper that cracked the problem was published by James Watson and Francis Crick in 1953. Using other people's data from X-ray crystallography (Appendix 2), they worked out that the strands of DNA, the main nucleic acid of the cell nucleus, were arranged in a regular double helix, two strands of DNA being wound round one another in a spiral staircase arrangement, with the nitrogenous bases providing the treads of the staircase and placed so that G (large) was always linking with C (small) from the opposite strand, and similarly A (large) with T (small) (Fig. 33.1). The links in this 'base pairing' are hydrogen bonds, which are similarly important in the context of protein structure, where they hold together the regular elements of secondary structure (Appendix 3).

DNA replication

To understand the significance of this discovery, which has literally transformed every corner of biology and medicine, one has to stop and think exactly what is required of the genetic material. Every time a cell divides, the genetic material has to divide too, and it must do so in such a way that it does not become diluted with each new generation. This means it has to replicate, producing a perfect copy of itself to pass on to the next generation. If we can separate the strands, the G-C and A-T rule for base pairing between strands means that each strand provides a perfect mould or template for making its opposite partner. All we require is an enzyme that knows the rules and will build up a new strand strictly following the base pairing. The enzyme DNA polymerase does this. The compulsory G-C and A-T base pairing, of course, immediately accounts for Chargaff's rule.

The double helix with its base pairing therefore explains how genetic information can be handed on faithfully from generation to generation at each cell division.

Self-test MCQs on Topic 33

1 Each nucleotide unit in a DNA strand is joined to the next through which of the following?

(a) Phosphate

(b) Sugar unit

(c) Nitrogenous base

(d) Amino acid.

2 The significance of the double helical structure of DNA is that

   (a) it neutralises the charge on the phosphate groups

   (b) it offers a way for daughter copies to be produced

   (c) it means that the genes take up less space in the nucleus

   (d) it locks the strands together so the genes are not degraded.

3 Chargaff's rules are explained by which of the following?

   (a) Nucleotide base pairing

   (b) Inter-species variation

   (c) Simple random statistics

   (d) Linear strand structure

Triplet codons

If replication were all that DNA could do it would be both interesting and impressive, but perhaps not much use. There is not much point in handing on a blueprint down the generations if no one understands how to execute it! DNA also somehow has to encode proteins' amino acid sequences. Since there are only four different bases in DNA, then, if there were a one-for-one code, e.g. G for alanine, T for glutamate, etc., we could only have four kinds of amino acids in our proteins. We have 20. Even with a two-letter code, AG, TC, AA, CG and so on, there are only $4 \times 4 = 16$ different possibilities, which is not quite enough. The molecular geneticists deduced, therefore, that there had to be (at least) a **triplet code,** i.e. a string of three bases coding for each amino acid. $4 \times 4 \times 4 = 64$, which is, of course, much more than enough.

The genetic code

Experiments in the 1960s gradually not only proved that there was indeed a triplet **genetic code** but also worked out exactly what it was. In fact, every 1 of the 64 three-base sequences (AGC, TGG, ACA, etc.), called **codons**, is used. Four of them are used as punctuation marks, three for STOP and one for START. This is necessary because the DNA is not in gene-sized pieces; each very long piece of DNA carries very many genes, one after another, and also a lot of intervening sequences. The reading machinery therefore needs to be told where each gene begins and ends. The

START codon AUG in fact does double duty. If it occurs at the beginning of a gene, then the reading machinery will come to warning signals in the DNA sequence to alert it to the fact that it is approaching the start of a new gene, and then it is prepared for the START codon. If it encounters the same codon in the middle of the gene sequence, it simply treats it as a codon for the amino acid methionine, one of the set of 20. In all, 61 of the 64 possible codons, therefore, specify individual amino acids. Only two amino acids have unique codons, methionine and tryptophan. The remaining 59 codons are divided between the other 18 amino acids. Some have two possible codons, some three, some four and some even six! This excess is termed redundancy (see Appendix 11).

Messenger RNA

It is not, in fact, the DNA itself that directly passes the coded message to the protein synthetic machinery. The DNA, as noted, is inside the cell nucleus. This is not where proteins are made. They are made out in the cytosol, and so the information has to be transported from one compartment of the cell to another. In this process, as we have just seen, the sequence of millions of bases in chromosomal DNA is interpreted and divided up into meaningful smaller chunks, individual genes corresponding to individual proteins. This process is called **transcription**, and it produces copies in the form of 'messenger RNA' (mRNA). RNA stands for ribonucleic acid and apart from the sheer size, the two key chemical differences between mRNA and DNA are that (1) the sugar units in RNA are ribose instead of deoxyribose and (2) instead of thymine (T), RNA contains **uracil** (U).

Transcription is carried out by an enzyme, RNA polymerase, that produces a complementary base-paired RNA strand to match the coding sequence in the DNA. The mRNA copies move across the nuclear membrane to the sites of protein synthesis in the cytosol. There can be many mRNA copies produced for a single DNA gene sequence, and this is what happens when a particular protein is required in large amounts. On the other hand, in contrast with the permanent and more or less indestructible DNA version, the mRNA copies are essentially transient and may be removed and broken down if the protein is no longer needed. Some proteins are needed constantly, and the corresponding genes are referred to as 'housekeeping genes'. Others are only needed for 'special occasions', and it is their messages that are likely to have only a temporary existence.

We have seen how single 'master copies' of the genetic information can be used to produce multiple copies in the form of mRNA. How are these multiple sets of instructions carried out and enacted in the form of proteins? This is the puzzle to solve in the next topic.

Self-test MCQs on Topic 34

1 Which of the following is true about the genetic code?

   (a) There are 20 codons for 20 amino acids.

   (b) There are 3 codons for each amino acid.

   (c) There are 64 codons for each gene.

   (d) There are 64 codons for 20 amino acids.

2 Redundancy in DNA coding means which one of the following?

   (a) That much of the DNA does not code for protein.

   (b) That there are several copies of each gene.

   (c) That most amino acids have more than one codon.

   (d) That there is a spare copy of each gene on the opposite strand.

3 The DNA reading machinery in the cell nucleus can identify the beginning of a new gene because

   (a) there is always a capital letter at the beginning

   (b) a new gene starts with a string of U's preceding the first codon

   (c) there is a specific start codon preceded by a warning signal

   (d) there is always a break in the base pairing at that point.

4 The sequence GGGUUAGTAC would not be found in DNA because

   (a) uridine is only found in RNA not DNA

   (b) triple repeats (GGG) are forbidden

   (c) the stop codon UAG would prevent translation

   (d) base-pairing partners A and T side by side would lock the structure.

Protein synthesis, ribosomes and tRNA

How is the genetic code put into practice?

We have shown in the last topic that a triplet code with four letters (DNA/RNA bases) to choose from is mathematically sufficient to encode 20 amino acids, but is it structurally sufficient? What is there about a set of three nucleotide bases that enables it to recognise and specify the very different chemical structure of a single amino acid? Likewise, how can very similar codons specify very different amino acids. How is it that AUG can specify methionine while, with only one base different in each case, AAG specifies lysine, GUG specifies valine, AUC specifies isoleucine? The answer to this puzzle lies, as so often, in the properties of a remarkable set of enzyme proteins and a matching set of RNA molecules called **transfer RNA** (tRNA), which together provide, in effect, an adaptor kit.

Transfer RNA

The tRNA molecules are only about 80 units long, i.e. they have about 80 bases, very much shorter than the typical gene, which is liable to have a coding sequence of 1000 bases or more (a protein 350 amino acids long will have a gene with 3×350 bases, i.e. 1050 bases). They are nevertheless long enough to coil up into a very well-defined 3-D structure, stabilised by a considerable amount of internal G-C and A-U base-pairing, where the RNA is folded back on itself in 'hairpin loops'. At either end of the structure are two features critical to the way tRNA works. At one end, the tRNA molecule presents the '**anticodon loop**'. This is a small loop of a few unpaired RNA bases in the middle of which is a triplet, a stretch of three bases,

Pain-Free Biochemistry   Paul C. Engel
© 2009 John Wiley & Sons, Ltd

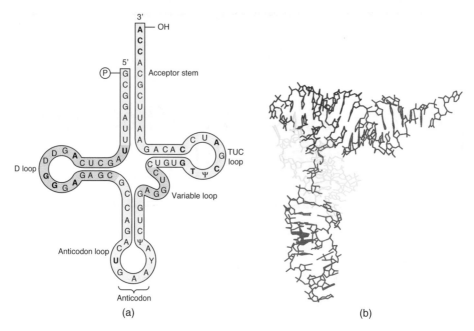

Figure 35.1 Structure of transfer RNA (tRNA). The 3-D structure (b) is shown schematically and squashed flat in (a) to emphasise base pairing, the acceptor stem and the anticodon loop. Reproduced from Pratt and Cornely (2004) *Essential Biochemistry*, John Wiley, New York. © 2004 John Wiley & Sons. Reprinted with permission of John Wiley & Sons, Inc.

that base pairs with one of the mRNA codons (Fig. 35.1). At the other end is a short stretch of unpaired RNA sequence, CCA, sticking out beyond a base-paired stem, the so-called **acceptor stem**. This is to receive the amino acid that corresponds to the codon that matches the anticodon loop at the other end of the molecule.

Attaching the right amino acid

This remarkable feat of molecular recognition seems almost too much to believe, but the thing that makes it all work is that, in addition, to complete the kit, each codon also has a corresponding enzyme, its **amino acyl tRNA synthetase**. There is a large set of these, just as there is a large set of different tRNAs. For each codon there is one specific tRNA that recognises it through the anticodon loop and one specific amino acyl tRNA synthetase that recognises the right tRNA molecule *and* the right amino acid corresponding to the codon. Since it is an enzyme, having achieved its highly specific job of double molecular recognition (tRNA and amino acid), it also has a chemical job to do and that is, in an ATP-driven reaction, to attach the amino acid via its carboxyl group to the ribose group of the terminal A in the acceptor stem.

So the faithful execution of the genetic code depends on the properties of a remarkable set of adaptor proteins. There is incidentally an intriguing chicken-and-egg question wrapped up in all this – well beyond our present scope, but tantalising to think about nonetheless: if proteins need nucleic acids to specify their amino acid sequence, and nucleic acids need proteins to carry out their coding mission, which actually came first, protein or nucleic acid, and in any case how could one work without the other?

Ribosomal protein synthesis

We still have not actually made any protein, but we now have the machinery that will allow us to select amino acids one by one to insert them into a growing protein chain. This process involves one final essential piece of equipment, the **ribosome**. Cells typically have numerous ribosomes scattered through the cytosol. Each one is a spool-like object that attaches to strands of mRNA and, beginning with the START codon, reads along, codon by codon, each time using the codon–anticodon recognition to bring up the right tRNA molecule carrying the correct amino acid. At any given moment, the ribosome is engaged with two of these charged tRNA molecules, and its catalytic machinery creates a peptide bond between the amino acids. This process, known as **translation**, requires GTP as part of the overall re-action mix. This is split to GDP with the formation of each peptide bond. As the ribosome rolls along the mRNA, a growing protein chain emerges.

Folding

The DNA and mRNA comprise linear information, a long string of nucleotide bases, and in the first instance what the ribosome produces is linear too, a long string of amino acids joined in a polypeptide chain. Working proteins, however, are not just waving strings of amino acids; as we have seen in earlier sections, they have very precise 3-D shapes. The final amazing feature of the process of protein synthesis, then, is that, essentially, the protein knows how to fold itself into the right final shape! Some proteins require a little shelter or a little molecular prompting to go in the right direction, but basically all the information is there in the amino acid sequence. Appendix 3 deals with the types of force between different parts of the protein structure that enable it to achieve this feat.

Self-test MCQs on Topic 35

1 Which of the following statements about protein synthesis is false?

(a) It takes place in the cell nucleus.

(b) It requires GTP.

(c) It requires a full set of amino acids.

(d) It requires ribosomes.

2 Which of the following statements about transfer RNA (tRNA) is false?

(a) A transfer RNA molecule recognises the right codon in mRNA.

(b) A transfer RNA molecule carries its amino acid into the translation machinery.

(c) A transfer RNA molecule recognises the correct gene for transcription.

(d) A transfer RNA molecule is about 80 nucleotide bases long.

3 If one of the 20 protein amino acids is missing, the ribosome will do which one of the following?

(a) Substitute the most similar amino acid.

(b) Skip the positions where that amino acid should go.

(c) Make a random substitution.

(d) Stall at the first codon for that amino acid.

TOPIC 36

Genetic differences and disease

Mutation

As we noted in Topic 32, genes account for both the sameness and the differences between different members of the same biological species. The sameness is ensured by the amazing power and fidelity of the machinery we have just described. The differences result because in reality even the best machinery is not perfect, and processes of such extreme complexity are bound to fail once in a while. If the machinery of DNA replication makes a mistake, this results in the wrong DNA sequence, i.e. a mutation. If this happens in the production of the germ cells, then the resulting individual produced from such a germ cell will carry the mutation, which will be faithfully copied each time a cell divides. Appendix 12 explains that there are mutations of different types. At one extreme, there are so-called 'silent' mutations that produce no change in the encoded protein. At the other extreme, there are mutations that may result in either no protein or an entirely jumbled, unrecognisable protein. In between, there are mutations that result in an altered protein that may or may not still be biologically active.

In the very long term, it is essential for living things that the genetic machinery allows a small but significant error rate. Without mutations it would not be possible for species to evolve and adapt to meet new environmental challenges. Allowing a small proportion of mistakes through allows them to be tried out to see whether by chance they offer an improvement, a selective advantage. The price, however, is that inevitably many of the mutations will be damaging, some only slightly, some disastrously.

Pain-Free Biochemistry   Paul C. Engel
© 2009 John Wiley & Sons, Ltd

Dominant and recessive

The outcome of a mutation in humans is complicated by the fact that for most genes we carry two copies, one inherited from each parent. Forty-four of our forty-six chromosomes are in pairs, the other two being the X and Y sex chromosomes. Most of the time, therefore, if a mutation results in a gene that encodes a defective protein, there is still a second copy of the gene, encoding a serviceable protein. There are still situations where we can only function properly with both copies of the gene intact, and in this case one bad copy results in disease. Such a mutation is referred to as **dominant** because its bad influence overcomes the efforts of the good copy. More commonly mutations are 'recessive', which means that we can get by with one good copy. A person with a **recessive mutation** is called a carrier. Carriers are apparently unaffected by the mutation themselves, but if they should have children with another individual who carries a defective copy of the same gene then there is a 25% chance that any child will inherit defective copies from *both* parents and therefore show the clinical symptoms of the deficiency. This is a situation that is especially likely to arise in communities with a tradition of 'consanguineous' marriages, e.g. between cousins.

Box 9   Cystic fibrosis

Among the more common genetic diseases is cystic fibrosis. This condition leads babies to have unusually salty sweat and lungs heavily congested with mucus. The latter symptom in the past used to lead on to infection, and CF patients typically died in childhood or in their teens. Paediatric chest wards would have nurses or parents patting children on their backs in an effort to dislodge all the unwanted mucus. Antibiotic treatment to control infection has extended these patients' life expectancy.

Until DNA methods became available, however, the nature of the mutation responsible for this child-killing disease was a mystery. Even with DNA methods, the task was like looking for a needle in a haystack. Researchers had to hunt through the DNA profiles of large numbers of individuals to find the telltale markers that were invariably present with CF and absent in totally unaffected individuals. In between these two categories were the 'carriers', individuals unaffected themselves but nevertheless carrying one defective copy of the gene. The problem is that, if you compare any two individuals' DNA, in addition to the one significant difference that you are looking for, there are also hundreds of other differences that are totally irrelevant to the search. Nevertheless, perseverance led to chromosome 7 and specifically to its long arm, and eventually the mutation was traced to a gene encoding a very large membrane protein, 1480 amino acids long. Once the sequence of this gene was determined, its resemblance to certain other genes for known proteins gave a big clue to what it was,* namely a 'transport protein' involved in moving chloride ion, Cl⁻ across membranes (see Topic 50).

Now that the gene has been identified it has been possible to examine it in large numbers of CF cases and discover that there are very many different possible disease mutations, i.e. either changes at different positions in the linear sequence and/or different changes at a particular position. Nevertheless, most of these are individually very rare. One particular mutation for some reason accounts for about three-quarters of all cases and is a very good example of the effects of mutation. This mutation is a deletion of three bases, resulting in an 'in-register' mutation (see Appendix 12) with just one amino acid fewer out of 1480. Nevertheless, this single change is sufficient to prevent the protein folding up properly. A tiny alteration at the DNA level in a single gene can result in death.

---

*Now that we know the amino acid sequences of many thousands of proteins from many hundreds of different biological species, it is obvious that different proteins fall into families more or less distantly related to one another. Proteins with different but similar functions tend to have very obviously similar amino acid sequences, indicating a common ancestry. Accordingly, we can argue the other way round, and, if we find a protein (or gene) of unknown function but with a similar sequence to one whose function is known, we can infer that the function of the new protein is likely to be similar to that of its cousin protein.

X-linkage

In terms of patterns of inheritance, an interesting situation arises in conditions that involve a gene situated on the X chromosome. Women have two copies of the X chromosome, but men have an X and a smaller Y. A mutation on the X chromosome could therefore be recessive in a woman, but if she passes on her defective chromosome to her son he has no second, good copy and therefore shows a disease that was hidden in his mother. Examples include classic **haemophilia**, the blood-clotting disease seen, for example, in a number of the descendants of Queen Victoria, who herself, being only a carrier, lived to a ripe and healthy old age. Another example is G6P dehydrogenase deficiency, which we met in Topic 29.

Patterns of disease

The clinical outcome of such mutations covers a wide spectrum. Some mutations may affect a gene that is so essential from the earliest stages of life that the individual cannot develop and survive even *in utero*. This might result in repeated miscarriages but will not result in sick live babies that need to be looked after. Other genes may be marginally less essential, perhaps because the mother's enzymes can cover for a deficiency in the foetus' biochemistry. In this case a baby may be carried to full term, resulting in a live birth, but possibly an infant that is clearly sick within days of birth as it struggles to manage without the maternal support systems. Beyond this are genetic disorders that may result in death at a later stage – **muscular dystrophy** in the teens and **Huntington's disease** in late middle age. Clearly a major factor affecting the prevalence of such conditions is whether they allow the affected individual to survive to sexual maturity and whether they affect his/her chances of finding a partner. There are also a considerable number of inherited conditions that are not lethal but cause discomfort or nuisance, say intolerance to certain foods, such as the widespread intolerance to the milk sugar, lactose, in Chinese populations.

Ethical issues

Increasing understanding of inherited diseases has brought with it better methods of early and reliable diagnosis, and this in turn confronts clinicians with ethical dilemmas, which they did not have had to face in the past. Many genetic diseases can be diagnosed at an early stage of pregnancy. Sometimes this may be a functional test, e.g. for an enzyme, but increasingly, in cases where one mutation predominates, it is possible to use sensitive DNA-based tests. In jurisdictions that allow termination of pregnancy, this kind of **antenatal screening** immediately raises, for parent and doctor alike, the issue of whether it is right knowingly to bring into the world a human being who will be disabled from the start and perhaps represent a severe burden. Equally it raises the contrary question of whether it can be right to terminate a potential new life. Routine screening in some cases, however, permits simple therapeutic intervention at birth. For example, **phenylketonuria** (PKU) is one of the most common enzyme defects, and, because it severely affects development of the nervous system, PKU is used to produce a high proportion of the inmates of mental institutions. However, diagnosis at birth (from a heel-prick bloodspot) now allows affected infants to be put on a special diet providing only low levels of the amino acid Phe (see also Box 6 and Topic 20). In the past, harsh Nature would have ruthlessly eliminated most such mutations after a generation or two. By our intervention we are, of course, tending to increase the 'genetic load' and storing up clinical problems for the future. Our increasing knowledge thus presents many difficult, controversial and painful issues for public health policy. At the same time, it allows intelligent **genetic counselling** of prospective parents and also, where there is an affected child, gives a better basis for medical care.

A particularly perplexing problem arises in the case of late-onset diseases like Huntington's disease, an invariably fatal, progressive neurological degenerative condition. If a parent is diagnosed, their children could already be in their teens or twenties. It is clearly possible to test them, but does such an individual want to know?

Box 10   Forensic DNA testing

As has already been mentioned, as human beings we all share a huge amount of genetic information that distinguishes us from other species. Even so, bearing in mind that we have about 25 000 genes, and that in between them there is a lot more DNA that does not encode proteins (or functional nucleic acids), even with a very small error rate, there are always going to be a large number of small, detailed differences between one individual and another. Each of us inherits DNA from both parents, so that there are two copies of most of the genome (the exception being the X and Y chromosomes). There are enough markers, however, to allow investigators to identify both parents with a high level of certainty.

Kinship relations can therefore be established with a high degree of confidence from DNA sequence. However, forensic DNA testing became widespread some years before whole genomes were being sequenced and has become an important tool in solving crimes (e.g. murder or rape cases) and in establishing parenthood (e.g. in paternity suits). This relies on four principal technical details:

1 So-called restriction enzymes (see Appendix 13) cut up DNA, but only cut at very specific sequences that may be six or even eight bases long. A precise sequence of that length does not occur very often, and so these enzymes cut DNA into a limited number of rather large fragments rather than a huge number of tiny fragments.

2 By using electrophoresis, DNA fragments of different size can readily be separated and seen, giving a pattern somewhat like a bar-code.

3 The technique of polymerase chain reaction (PCR) now allows us to take tiny amounts of DNA, such as can be recovered from a hair follicle, a cheek swab, a fingerprint, a cigarette stub, etc. and amplify so that there can be much larger amounts of identical DNA for analysis.

4 It was realised that our DNA has certain non-coding regions that contain repetitive stretches, where the same run of bases occurs over and over again. This leads to a lot of variability between individuals – one person might have 20 repeats and another might have 35.

Self-test MCQs on Topic 36

1 Consider the following four statements:

(1) A mutation can alter a gene without altering the encoded protein at all.

(2) A mutation can alter a protein without altering its function.

(3) A mutation can alter the length of a protein.

(4) a mutation can alter a stretch of amino acid sequence.

Choose your answer below:

(a) All correct except 4.

(b) Only 2 and 3 are correct.

(c) None are correct.

(d) All four are correct.

2 A dominant mutation is which one of the following?

(a) A mutation that improves the function of the affected gene.

(b) A mutation that leads to abnormally aggressive behaviour in males.

(c) A mutation that destroys function even if it affects only one gene copy.

(d) A mutation that has to be handed on by the mother to be effective.

(e) A mutation that has been passed on by both parents.

3 Which of the following is true about genetic defects?

(a) Genetic defects can only be diagnosed after birth from the protein or loss of function.

(b) Genetic defects can be assessed at the DNA level only after the age of 2 when a sufficient sample can be taken.

(c) Detection of genetic defects is only of academic interest since damage is irreversible.

(d) Detection of genetic defects either before or at birth assists therapy and genetic counselling.

Genetic variability: drug metabolism and disease susceptibility

Gene variability

For any human gene, there is what we would regard as the standard correct DNA sequence and corresponding protein amino acid sequence. As we saw in the previous topic, there are a variety of mistakes in these sequences that show up as recognised genetic diseases. Also, however, there are a host of minor variations that do not cause such obvious disturbance. If we were to take the same gene from a thousand individuals and determine the DNA sequence, we should certainly find a number of positions in the sequence where there was more than one possibility. Any individual, therefore, is liable to show minor departures from the standard sequences. These departures are known as genetic **polymorphisms**. The fact that they are not classified as diseases does not necessarily mean that they are all functionally neutral (i.e. without any effect at all). It is increasingly being realised that gene polymorphisms are clinically important in at least two different ways.

Drug metabolism

First of all there is the question of how we handle drugs. In general, the dosage of a drug that a doctor prescribes is related to body weight: you do not give as many tablets to a 9-year-old, 35 kg child as to his 80 kg father. On the other hand, adults of similar size would tend to be given the same dose. That seems to make sense, but only if their bodies deal with the drug in the same way. Over time, we excrete drugs, and more often than not chemically alter them first. Our body responds to foreign compounds as poisons and does its best to render them harmless and then get rid of them. This depends on the activities of various enzymes, and if there are variations in these activities in the human population, then one person may clear a drug much faster than another. If so, giving them the same dosage because they are of similar weight may not after all be appropriate. This realisation and the possibility in recent years of obtaining DNA sequence information for an individual relatively easily, quickly and cheaply have led to the study of **pharmacogenomics**. This is still in its infancy, but in years to come we may expect drug regimes to be much more closely tailored to what we know about the specific body chemistry of an individual patient.

Genetic risk factors

The second area of growing importance is that of susceptibility to disease. It has long been known that some common diseases run in families, although it was not known why. For the same reason, known disease risk factors carry very different levels of risk for different individuals. For example, in Topic 19 we met LDL, one of the proteins associated with transport of cholesterol. This has to dock with a specific protein receptor, the LDL receptor, at cell surfaces. A variety of mutations in the gene for the LDL receptor cause familial hypercholesterolaemia; affected individuals have a much enhanced risk of heart disease. Detailed study of the human **genome** is likely to throw up many more correlations of this kind for various kinds of cancer and other diseases. It has recently been found that a polymorphism related to survival of mediaeval plague epidemics may also be responsible for certain individuals' remarkable resistance to the HIV infection that causes AIDS!

Self-test MCQ on Topic 37

Pharmacogenomics refers to which of the following?

(a) The study of patients' differing genetic susceptibility to drugs.

(b) The study of genetic damage through long-term exposure to drugs.

(c) The study of cost–benefit aspects of drug therapy.

(d) The study of multiple drug resistance in bacteria.

TOPIC 38

Mutagens, radiation and ageing

Harmful DNA damage

We have already discussed the existence of a low error rate in the processes of DNA replication, leading to naturally occurring mutations. These error rates can be very greatly enhanced by both physical and chemical influences in our environment. Physical damage can be caused either by ultra-violet radiation, e.g. from **sunlight**, or by **X-rays** or ionising radiation emitted by **radioactive materials** (Chemistry X). Chemical damage to DNA can be caused by a variety of chemical substances such as the tar in cigarette **smoke** and various **dyes**, now banned, that used to be added to foods, supposedly to make them look more appealing.

These damaging effects on DNA can be seen as a double-edged sword. Most obviously they account for the link between smoking and **lung cancer**, or between basteing bodies on sun-drenched beaches and **skin cancer**, and also for the concern for stringent control over food additives and safety precautions in the chemical industry. In all these cases, one is not concerned about mutations that will lead to an inherited mutation but about mutations that will cause one among millions of cells in a child or an adult to go wrong and then multiply out of control. In case of penetrating radiation, there is also the risk that the germ cells that will lead to a new embryo may be damaged. This was responsible for the increase in **genetic defects** following the atomic bomb attacks on Hiroshima and Nagasaki and likewise the disastrous accident at the nuclear power plant at Chernobyl. Closer to routine health care, it calls for responsible precautions in **radiography**.

Pain-Free Biochemistry   Paul C. Engel
© 2009 John Wiley & Sons, Ltd

Useful DNA damage

Also, however, we can put DNA's susceptibility to damage to practical, targeted use. Thus, both ultra-violet light and gamma radiation are used for **sterilisation** of surfaces and objects that need to be microbiologically clean. It is not only human cells that depend on intact DNA to tell them what to do! Also, while radiation may sometimes be a cause of cancer, it can be used too in the effort to cure cancer. **Radiotherapy** is a regular part of the oncologist's armoury either following or instead of surgery. The intention here is, of course, not to cause a few mutations but rather, by massive doses of focused radiation, to inflict massive, fatal damage on the rogue cells.

Another process in which mutations in DNA are now thought to play a central role is **ageing**. It seems that as time goes on our cells accumulate more and more mutations. When these are relatively few, our systems can compensate and repair. Over time, however, the system is overwhelmed and becomes less and less able to cope.

Self-test MCQ on Topic 38

Which of the following damage DNA?

(a) Ultra-violet light

(b) X-rays

(c) Cigarette tar

(d) Both b and c but not a

(e) All three: a, b and c

Switching genes on and off: development, tissue specificity, adaptation and tolerance

The genome

Through a huge collaborative international scientific effort, the human genome was sequenced in its entirety a few years ago, i.e. the detailed sequence of bases was established for all the genes and also the bits of DNA in between the actual coding sequences. Since then, there has been a certain amount of tidying up and editing, as a result of which it seems to be accepted now that we have about 25 000 different genes. Scientists were rather surprised that the number was as small as that, particularly as it has emerged that some other creatures have considerably more, and other living things that we regard as lower down the evolutionary scale have as many. However, in all probability you would see 25 000 as quite a large number, and when you think in terms of an individual cell having to express those genes to turn out 25 000 different proteins, it seems like a considerable production challenge!

Tissue specialisation

In fact, our individual cells do not express 25 000 genes. Depending on the tissue, particular types of cell might express a larger or smaller fraction of that large total, but, in general, because cells are specialised, they need only a subset of all the

Pain-Free Biochemistry   Paul C. Engel
© 2009 John Wiley & Sons, Ltd

possible proteins and they do not waste time and energy making ones that they do not need. There are a certain number of so-called '**housekeeping genes**' that are needed most of the time in virtually every tissue, e.g. enzymes of glycolysis, but we might not expect all the proteins involved in muscle contraction to be expressed in the brain or the haemoglobin that fills the red cell to be abundant in the eye. This general issue was considered earlier in Box 8 (Topic 26).

Development and physiological adaptation

There is also another dimension to this – time. Many of our tissues are called upon to do different things at different times. There are all the changes as we gradually grow up. There are responses to environment or nutrition. There are also responses to injury and to major physiological changes such as pregnancy, lactation and weaning. We have already noted back in Topic 28, for example, that the mammary gland has a sudden demand for reducing power in the form of NADPH for biosynthesis at the onset of lactation. In preparation for this there is a big increase in the levels of the enzymes of the PPP.

Pharmacological tolerance

Our bodies are also able to respond to the non-standard challenges we present. Alcohol, pain-killing drugs, toxic substances, etc. will all produce very different degrees of response depending on whether the body is presented with a one-off challenge or a regular encounter. This is because regular treatment changes the levels of both the proteins in the body that respond to these substances (e.g. pain receptors) and of the proteins that work to eliminate them. This clearly has important implications for managing long-term drug dosages.

Gene switches

All this implies that genes, or sets of genes, are able to be switched off or on. Many hormones (see also Topic 41) act in this way, turning on the expression of sets of genes in particular target tissues. The detailed way in which this works is well beyond our scope in this book, but in brief there are numerous proteins that can interact directly with DNA, and are described as activators or **repressors** of transcription. The various switch mechanisms work by controlling the interaction of such proteins with their DNA targets, e.g. by clamping onto the DNA, a repressor will prevent transcription of a neighbouring section of the genome; any modification that lifts the clamp off the DNA will therefore switch on transcription.

Self-test MCQ on Topic 39

Which of the following statements is true regarding gene expression?

(a) About 25 000 genes are expressed in a human cell but at any one time most of them are inhibited.

(b) There are about 25 000 genes in a human cell but at any one time only a small fraction of them are expressed.

(c) There are about 25 000 genes simultaneously expressed in a human cell, accounting for the extraordinary metabolic flexibility.

(d) There are about 25 000 genes expressed in a human cell, but the majority of the proteins are immediately and selectively degraded.

DNA and protein synthesis as targets: chemotherapy, antibiotics, etc.

Is targeting DNA or protein synthesis a suicide mission?

We have already noted in Topic 38 that the central importance of DNA as the controlling substance in a cell makes it a very obvious target if there are cells that we want to destroy. In the medical context, in practice this means (1) cells of invading organisms and (2) our own rogue cells in cancerous growths. As well as physical assault with radiation, we have more selective chemical tools to deploy in this task. This involves the inevitable issue that surrounds any drug development, of how to achieve the objective without poisoning ourselves (or rather our patients!) in the process. Often, in attacking other, harmful organisms, we try to target a process that is essential to them but is not part of our own body chemistry (as, for example, with penicillin, which attacks the formation of bacterial cell walls – humans do not have cell walls, only cell membranes). In attacking the reactions surrounding DNA, RNA and protein synthesis, however, we are aiming at processes that are essential to every cell. This might seem to be a dangerous, if not impossible, route to pursue! Nevertheless, it is a route that medicinal chemists have followed. How is this possible?

Attacking protein synthesis in bacteria

Fortunately for us, even though many of the processes and reactions are identical or closely similar across many organisms, the proteins and other machinery, e.g.

ribosomes, are not identical. What we want is a compound that will block the machinery in a bacterium (or other infectious agent) without blocking the corresponding machinery in man. In the development of anti-bacterial agents, we have had a big helping hand from Nature. Sir Alexander Fleming first noticed that a fungus (*Penicillium*) could enhance its chances by killing bacteria in its vicinity. The active agent in this case, penicillin, does not attack DNA or protein synthesis. Rather, as mentioned above, it attacks an enzyme required for building the bacterial cell wall. The discovery of penicillin, however, not only led to the development of a variety of chemically modified penicillin drugs, but also prompted the search for other naturally occurring **antibiotics**. The next 'wonder drug', **streptomycin**, works by specifically inhibiting bacterial protein synthesis, and we now know that the reason it can work so selectively without harming the host is that bacterial ribosomes have a different structure (and size) from those of higher organisms. Therefore, chemicals that target the bacterial ribosome do not have the same affinity for ribosomes in humans and so do not poison us.

Antibiotic resistance

Streptomyces, the organism that makes streptomycin, has turned out to be the antibiotic equivalent of a gold mine! Various *Streptomyces* species produce **chloramphenicol, erythromycin, tetracycline**, etc. What is rather remarkable about this is that *Streptomyces* is not a fungus like *Penicillium*; it is a bacterium! So, of course, it too has bacterial ribosomes. How, then, does it escape its own poisons?

Nature has not only developed antibiotics as part of the armoury in the warfare between different organisms; it has also provided the organisms that make some of these compounds with their own defence – antibiotic resistance. This can work in various ways, but often the organism that makes the antibiotic has an extra enzyme that inactivates the dangerous compound. Most of the active antibiotic will in any case be put out into the surroundings, but any stray lethal molecules that sneak back and try to kill the wrong bug are destroyed. For example, organisms that are resistant to chloramphenicol have an enzyme that adds an extra acetyl group (CH_3CO-) to the antibiotic molecule, rendering it harmless.

For a couple of decades medical science seemed to be on the verge of eliminating the killer bacterial infections, but the bacteria have fought back. The enzymes responsible for antibiotic resistance in organisms like *Streptomyces* are encoded by their own genes, and what we have discovered in recent decades is that there are mechanisms that allow genes to jump species, so that resistance can be transferred from one kind of organism to another. Statistically, this might be a very rare event, but bacteria multiply very fast, and once they get growing there will be many millions even in a millilitre of liquid. Even if the transfer of resistance happens only to one cell in a thousand million, as soon as the antibiotic is there, that one cell in a thousand million has a massive advantage over the other 999 999 999, which are all likely to be killed. With the coast clear, our one resistant cell can now multiply and produce a multitude of identical, resistant cells. This is natural selection at work. It is effective and in bacteria it can be surprisingly fast.

By being careless in our overuse of antibiotics and in the way we handle them, we have over a period of time favoured the survival and multiplication of just those bacteria that are most dangerous, the ones that carry resistance genes! So now we have a continuing race between pharmaceutical chemists trying to discover or synthesise novel antibiotics and the ingenuity of the bacteria in coming up with new defences. Diseases like tuberculosis, which were thought to be almost eradicated in developed countries, are now coming back with a vengeance. It has become clear too that, even when the drug is something entirely new produced by synthetic chemistry rather than biology, it is only a matter of time before bacteria come up with a new defence, using the power of mutation to innovate (see, e.g. Topic 50)!

This is an issue with huge implications for nursing. Patients are becoming scared of hospitals for fear that medical attention might kill them! Methacillin-resistant *Staphylococcus aureus* (**MRSA**), *Clostridium difficile*, lethal strains of *E. coli* are constantly in the news. The people most likely to spread these killers from patient to patient are doctors and nurses. Up till the 1940s, hospital staff were obsessive over hygiene; there were no antibiotics. Today we have become complacent and careless and the bacteria are back!

Targeting our own cells

When it comes to dealing with cancer, we have a much tougher problem because these are our own cells out of control. A drug that targeted ribosomes in the cancer cell would also act on ribosomes in all the other cells of the body because they are the same kind of ribosomes. Since most cells need to make proteins, such a drug would be toxic to nearly all our cells. We achieve selectivity in a different way, by targeting DNA synthesis. Cells need to make DNA only when they are dividing, and so with a short, sharp dose of **chemotherapy** we target mainly the rapidly dividing cells, such as those of an aggressive cancer. Unfortunately, these are not the only rapidly dividing cells in our bodies. The mucosal linings in our digestive and respiratory tracts and the cells in our bone marrow are also constantly dividing. This means that unless the drug can be delivered or targeted directly to the cancer, there is bound to be 'collateral damage'. This is why chemotherapy has all its well-known unpleasant side-effects.

How do we target DNA synthesis? Two typical drugs (Fig. 40.1) are **5-fluorouracil** and **methotrexate**, and both of them target the supply of thymine nucleotides. As we saw in Topics 33 and 34, thymine is one of the four DNA bases. RNA has uracil instead of thymine, and the difference between them is that thymine has an extra CH_3 group (Fig. 40.2). Uracil thus supplies building blocks for RNA, but it is also a precursor in the pathway to make the thymine building blocks for DNA. Uracil is converted into deoxyuridine monophosphate, and in the next step the CH_3 group is added to make deoxythymidine monophosphate (Fig. 40.3). The

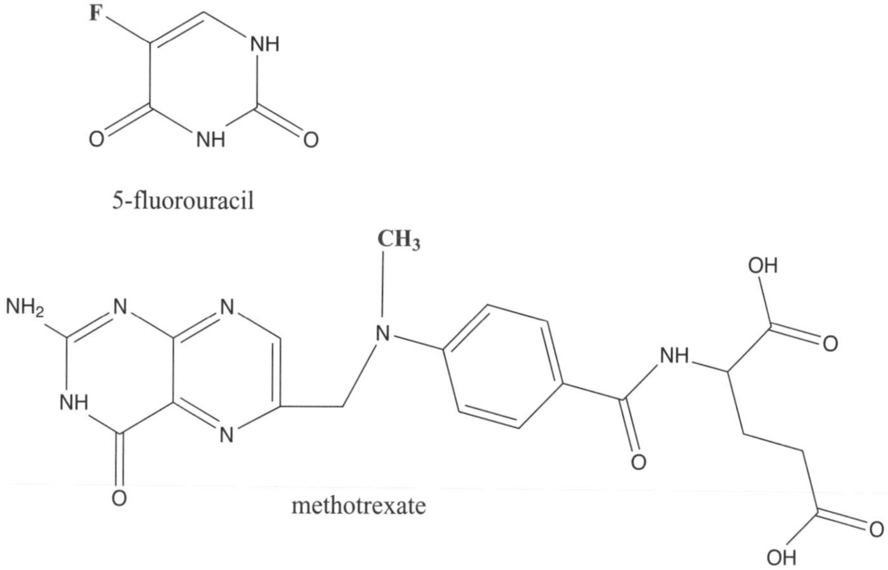

5-fluorouracil

methotrexate

Figure 40.1   Drugs that target DNA synthesis.

donor for this process is tetrahydrofolate, a cofactor derived from folic acid. Both the drugs we have mentioned block this reaction in different ways. 5-Fluorouracil is like a wolf in sheep's clothing: it fools the enzymes, which treat it as if it was uracil, and so it gets converted into 5-fluorodeoxyuridine monophosphate, which steps up for the next enzyme step. The next enzyme, however, is in for a surprise because this impostor substrate reacts with the enzyme protein in an irreversible way so that it is permanently blocked and inactive!

Methotrexate acts one step away by blocking the supply of tetrahydrofolate; it is a very similar chemical compound to folic acid and again exercises its action in effect by fooling the enzyme, occupying the place that should be taken by the genuine compound (see Appendix 14).

These drugs are two among many that have been developed to target DNA synthesis, and sometimes therapeutic regimes employ cocktails of several of these

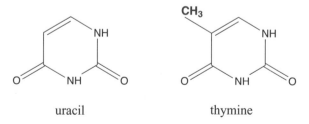

uracil   thymine

Figure 40.2   Thymine and uracil. Both these nucleic acid bases form base pairs with adenine, but DNA exclusively uses T rather than U, and RNA uses U rather than T.

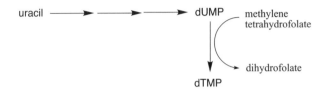

Figure 40.3 Metabolic processing of uracil to make deoxyTMP for DNA synthesis on the way to deoxyTTP, one of the four building bocks for DNA synthesis. This pathway is blocked by the anti-cancer drugs methotrexate and 5-fluorouracil.

drugs. In spite of their side effects, these agents have dramatically improved the chances of patients with some types of cancer, testicular cancer for example. As already mentioned, the only way to avoid the side effects would be to deliver the drug more specifically to its target. One possible way to do this that has been explored is to use an antibody that can specifically recognise antigens on the surface of the cancer cell as a delivery mechanism for an attached drug molecule. In some cases, however, the antibody itself may be the best drug! **Herceptin** (for some types of breast cancer) is a case in point. Note: This does NOT target DNA synthesis, but it does very effectively target the right cells.

Self-test MCQs on Topic 40

1 Which of the following is true regarding radiotherapy?

(a) It is difficult to achieve DNA damage because of DNA's intrinsic stability.

(b) It is important to kill the cells without damaging their DNA.

(c) It is difficult to avoid DNA damage to other cells than the targeted ones.

(d) It is impossible to avoid unpleasant toxic symptoms resulting from DNA fragments.

2 Which of the following is true regarding anti-bacterial action?

(a) It is easy to target bacterial protein synthesis because it is identical to our own which is well understood.

(b) It is risky, on the whole, to target bacterial protein synthesis because unavoidably we simultaneously attack our own.

(c) It may be possible in the future to target bacterial protein synthesis but it is not yet well understood.

(d) It is useful to target bacterial protein synthesis because there are key differences from our own allowing for selective inhibition.

3 Antibiotic resistance refers to which of the following?

(a) Many pathogenic bacteria have evolved or acquired the ability to get rid of antibiotics.

(b) Many patients are unable to tolerate antibiotic substances.

(c) Many antibiotic molecules are resistant to biological degradation.

(d) Many clinicians are reluctant to prescribe antibiotics because of side effects.

SECTION 5

Physiological Systems and Clinical Issues

TOPIC 41

Hormones and second messengers

We are complex multicellular creatures, and our multitude of cells is, of course, of many different types, making up different tissues and organs in different parts of our bodies. This can only work and offer us an advantage over, say, an amoeba, if these many kinds of cell and organ can communicate with one another so that they can cooperate efficiently. The experiments of physiologists over the past hundred years or so have revealed two ways of doing this. One that is very rapid is delivering electric pulses via the nervous system. The other, slightly slower but longer lasting and more pervasive, is the effect of hormones secreted by various endocrine glands – the thyroid, the adrenals, the gonads, the parathyroid, the pancreas and above all the chief controller, the pituitary. But what are hormones? As they were isolated, one by one, during the last century, it became obvious that they are a hugely diverse group of biochemical compounds. There is nothing structurally in common between, say, a sex hormone like the steroid progesterone, with quite a small molecule ($M_r = 314$), and the pancreatic hormone insulin, which is a protein (51 amino acids, $M_r = 5734$). So the name 'hormone' links them, not because of what they are chemically, but because of what they do physiologically in a very general sense. They are all chemical messengers. They are put out by one organ to tell another organ what to do or what not to do.

Frequently our two modes of signalling work together in harness: our sensory processes result in a nervous signal, which stimulates a gland to release a hormone that then, travelling via the bloodstream, can act on multiple target organs round the body (as we shall see, for example, in Topic 43 in the case of adrenaline). How do hormones work? Can we, in fact, generalise, bearing in mind that we have just noted that they are so chemically diverse? Surprisingly, perhaps, we can. A considerable number of hormones *do not enter the cells/organs* they affect! How can this be?

Pain-Free Biochemistry   Paul C. Engel
© 2009 John Wiley & Sons, Ltd

In rather the same way that the postman can stir up a household without coming through the door! At the surface of the cell, sitting in the cell membrane, are special proteins that are **hormone receptors**. They are specifically shaped to recognise the hormone and allow it to dock, and in effect they tell the cell that the postman has arrived! However, they cannot shout and they are stuck at the front door (i.e. in the cell membrane) and so what they have to do is switch on a messaging system inside the cell, something that they are well placed to do, straddling the membrane as they do. This, then, gives rise to the concept of the **second messenger**. The first messenger is the hormone, but because the first messenger does not enter the cell, there has to be a second messenger to relay the message from the membrane to the rest of the cell.

The way in which this operates was first worked out in the 1960s for one hormone in particular, namely **adrenaline**, but the ideas and mechanisms that emerged turned out to have far wider applicability, explaining the mode of action also of many other hormones (each with its own membrane receptors). This has led on to a whole new branch of biochemistry, the study of 'cell signalling'. In the case of adrenaline and a considerable number of other hormones, the docking of the hormone on its receptor triggers the activity of an enzyme called **adenyl cyclase**. This enzyme takes ATP and converts it into '**cyclic AMP**'. We have already met ordinary AMP, with its phosphate group attached to the $5'$-position of the ribose ring of adenosine (see Appendix 5). Cyclic AMP (or cAMP) is so called because its phosphate is linked to both the $5'$ and the $3'$ position. This makes it a different compound, with a different shape and activity. What cAMP does in the context of hormone action and signalling is that it switches on an enzyme called protein kinase c. A kinase is an enzyme that uses ATP to put phosphate on another molecule and this one is called *protein* kinase because it attaches the phosphate to protein molecules; not all protein molecules, but certain ones tailored to be receptive to the attentions of the protein kinase. The point of this addition of phosphate to the protein (phosphorylation) is that this in turn acts as an on switch for that protein.

What emerged from the adrenaline study was a **cascade** of these enzymes, one switching on the next, which then switched on the next and so on. What is the point of such a complicated sequence? It allows a rapid and massive amplification of the initial signal. A single enzyme molecule can act on many other molecules in a short time. If each of those molecules is itself an enzyme and can act on many of the next lot of target enzyme molecules, you can rapidly get a huge multiplicative effect.

With switches and taps, it is just as important to be able to switch them off as on, and this is the beauty of the phosphorylation mechanism. Just as you can switch activity on by adding a phosphate, you can equally reverse the effect by removing it, and so these signalling pathways invariably match each kinase (which puts phosphate on) with a **phosphatase**, which takes it off again. All of them need to be tightly controlled, but mechanisms of this sort now turn out to control a vast number of biological processes, not only the response to adrenaline, but also cell division, egg fertilisation and so on.

The need for an off switch applies not only to protein phosphorylation but also to the initial formation of cAMP. If the overall process has to be switched off, it is essential to remove the second messenger; otherwise it will constantly be switched back on again. So alongside adenyl cyclase there is a second enzyme, cAMP phosphodiesterase, which splits cAMP to give ordinary 5′ AMP, which does not exert the same effect. This enzyme is the site of action of various stimulatory drugs (see Box 11).

Box 11   Caffeine

Caffeine is one of a group of substances called methylxanthines. They are closely similar compounds to adenine and guanine, which are building blocks in the nucleotides AMP, ADP, ATP and GMP, GDP and GTP. Caffeine is the main stimulant substance in coffee but similar compounds, theophylline and theobromine, are found in cocoa and tea. All of them promote wakefulness. Caffeine is the stimulant in such drinks as Red Bull. One of the sites of action of these substances is the enzyme **cAMP phosphodiesterase**, where they act as inhibitors (see Appendix 13). This effectively keeps the systems that are activated by cAMP switched on. It is now known also that these compounds have another site of action in the brain and elsewhere. In these locations there are inhibitory adenosine receptors, upon which caffeine acts as an antagonist.

Self-test MCQs on Topic 41

1 Hormones are which of the following?
   (a) Second messengers delivering regulatory signals to the site of action.
   (b) Receptors for neuroregulation of different organs.
   (c) Carriers of messages from endocrine glands to receptors on target organs.
   (d) Enzymes switched on in response to a second messenger.

2 Cyclic AMP is which of the following?
   (a) A second messenger.
   (b) The product of protein kinase action.
   (c) A hormone.
   (d) The term for cascade amplification.

TOPIC 42

Switching enzymes on and off: coarse and fine control

Our bodies have a series of ways of controlling enzyme activities, acting on different timescales to cope with different challenges. In a sense, at one end of the scale you might argue that the slowest control measure of all, affecting the whole species over millions of years, is the total elimination of a gene or set of genes. For instance, evolution appears in effect to have taken a calculated risk with the enzymes to make the various vitamin molecules that we need for our enzyme cofactors. We do not need to make these compounds so long as we can rely on eating some other living thing that can! That is a rather long-term and irreversible sort of control. More on our own timescale, there is the switching on and off of genes or suites of genes. This may be in response to a programmed developmental change or it might be a specific response to a changed physiological situation such as pregnancy or starvation. This is termed **coarse control** because it is rather slow both to take effect and to reverse, like changing the direction of an ocean liner. Changes in enzyme or other protein levels by this kind of control are likely to occur over many hours and sometimes days or weeks.

Clearly responses on that timescale are adequate as a response to a long-term metabolic challenge like pregnancy but not for an immediate emergency – say the need to run away from danger. Our ancestors on the African plains would not have lasted long if they had to make some more glycogen phosphorylase every time they were chased by a lion! So for this kind of situation we have **fine control**, acting not at the level of the gene but at the level of the individual protein molecule. We have already seen in the last topic and earlier in Topic 30 that one way to do this is by covalent modification of the protein, i.e. by a chemical change. In the case of attaching phosphate, this is set up to be a reversible control – the phosphate can be removed. There are other processes like the switching on of blood clotting (Topic 54)

Pain-Free Biochemistry   Paul C. Engel
© 2009 John Wiley & Sons, Ltd

in which the relevant protein is swung into action by proteolytic action, i.e. splitting the polypeptide chain and perhaps removing a piece. This is an irreversible change, like a weapon that is designed to be fired only once.

There is, however, another way of exercising fine control, which offers a way of modulating activity instantaneously and constantly. This is by a process called **allosteric regulation**. The name implies ('allo' = different; 'steric' refers to shape) that an enzyme can be regulated by substances that look nothing like the actual substrate. This was a surprise when the process was first discovered because everyone was accustomed to the idea that enzymes have active sites that very selectively recognise the right substrate and nothing else. However, we have come to realise that some enzyme molecules are very flexible and can readily change their shape, and also that, as well as active sites for their substrate molecules, they may have **auxiliary sites** for other molecules. If a regulatory substance, possibly a key metabolite such as ATP, attaches itself (non-covalently and temporarily) to such an auxiliary site, it may provoke a shape change in the protein so that the **active site** is distorted. In principle such 'allosteric' effects may either activate the enzyme or inhibit it.

The set of genes that we are equipped with has been selected and fine-tuned through millions of years of evolution to keep us reasonably versatile and prepared for likely events. On the other hand, making proteins that are not really needed is an expensive business. Since only the liver is supposed to make ketone bodies (Topic 21), there is no point in having the specific enzymes for that purpose in brain cells or cardiac muscle. Equally, even though the mammary gland makes milk fat, it does not do so all the time, and there is no sense in making the necessary enzymes when they are not needed. Often an enzyme activity in the wrong place at the wrong time may do more harm than good. In the example, just considered in the last topic, of the ability of kinases and phosphatases to put phosphate onto proteins and remove it again (to switch them on and off), if the phosphatases and kinases were active at the same time, they would fight one another to a standstill, achieving nothing. In fact, it would be much worse than merely pointless because the kinases use ATP, which is not recovered in the phosphatase reaction. This would be another example of a 'futile cycle', as discussed in Topic 31, and it would quickly squander the cell's hard-earned ATP.

Self-test MCQ on Topic 42

Which of the following methods may a cell use to make sure a particular enzyme activity is missing?

(a) Switching off the gene that encodes the enzyme.

(b) Chemically altering the enzyme protein, e.g. by phosphorylation.

(c) Changing the enzyme molecule's shape via a regulatory site.

(d) All of the above.

TOPIC 43

Insulin, glucagon and adrenaline

In Topic 41, we considered in general terms the way in which hormones act as chemical messengers, transmitting information from one tissue to another. In this topic, we are going to look at three of the most important hormones and the way in which they control how we handle major foodstuffs in three major tissues: muscle, liver and adipose tissue (which, between them, make up a high percentage of total body mass). The three hormones are insulin, glucagon and adrenaline, two of them produced by the pancreas and the third, as its name implies, by the adrenal medulla. Their effects, explained below, are summarised in Table 43.1.

Insulin

Insulin is produced by the β-cells in the islets of Langerhans of the pancreas and can be seen as a response to the well-fed state. In that state there will be a good supply of glucose, amino acids and triglyceride coming from the small intestine. Insulin has a powerful effect on most tissues, and it affects the handling of most major nutrients. Although these effects are multiple, they are actually easy to learn because they make obvious sense. If you do not remember them you can predict them and be confident of being right:

- Starting with glycogen, if glucose is plentiful, do we need to break down glycogen? No. Accordingly we find that insulin switches off glycogen breakdown in both liver and muscle. On the other hand, it would make good sense to store away some of the sugar for a rainy day, and so insulin stimulates the synthesis of new glycogen in liver and muscle. Two opposing effects on two opposing processes.

Table 43.1 Summary of main metabolic effects of insulin, glucagon and adrenaline

Hormone:	Insulin	Glucagon	Adrenaline
Timescale	Long term	Long term	Short term
Sites of action	Throughout body	Liver	Muscle, liver and adipose
Glycogen breakdown	↓	↑	↑
Glycogen synthesis	↑	↓	↓
Glycolysis	↑	↓	
Gluconeogenesis	↓	↑	
Glucose uptake	Except brain and liver ↑		
Fat deposition	↑	↓	↓
Fat mobilisation	↓	↑	↑
β-oxidation	↓	↑	

- Gluconeogenesis, making new glucose from other precursors, is not an activity of muscle or adipose tissue but is an important function of the liver (and kidneys). If glucose is plentiful, however, we do not need to be diverting amino acids and so on in that direction, and so insulin switches off gluconeogenesis.

- Glycolysis is up-regulated in all three tissues by insulin. Since glucose is plentiful for the time being, it is not essential to spare it for the exclusive use of brain and nervous tissue.

- Turning to glucose, its utilisation in external tissues is limited by controlled uptake from the blood, and so insulin massively stimulates this uptake into muscle and adipose tissue via its effect on insulin receptors in the cell membranes of these tissues.

- The fed state is also an opportunity to lay down storage triglycerides in adipose tissue. Insulin therefore switches on lipoprotein lipase, which downloads fatty acids from chylomicrons; it switches on the enzymes that resynthesise triglyceride inside the adipocytes; it switches off hormone-sensitive lipase, the one that releases fatty acids into the blood when the stores are mobilised.

Thus, insulin switches off all the emergency supplies, it encourages catabolism of the plentiful food-derived carbohydrate supplies while they last and it also stimulates the laying down of new reserves, saved from the surplus, of both glycogen and fat.

Glucagon

Times are not always so good, however, and we need to be able to respond appropriately also to the hard times, i.e. to fasting. A first step is to switch off the supply of insulin from the β-cells, since it is no longer helpful, but this alone is not enough. The pancreas, however, is still in control and switches to the cells next door, the α-cells, also in the islets of Langerhans, which secrete a different protein hormone, glucagon. This happens in response to falling blood glucose levels, and the job of glucagon is to get the blood glucose back up and maintain the necessary level as long as possible. In this respect, insulin and glucagon behave as 'opposite' hormones. However, there is one crucial difference: insulin, as we have seen, acts all over the body, on many different kinds of tissue; glucagon works exclusively on the liver.

Thinking about the role of the liver in body metabolism, we should now be able to predict what glucagon needs to do:

- **Glycogen** (not to be confused with the similar-looking name of the hormone!). Do we need to make it or break it down? Since we need more glucose we should convert glycogen into glucose, not vice versa. Therefore, glycogen synthesis is switched off and glycogen breakdown is switched on.

- **Glycolysis**. The liver is now being called upon to do its bodily duty and put glucose out into the bloodstream. It would therefore be most unseemly for the liver to be seen with its own snout in the trough. Glucagon has to switch liver glycolysis off.

- **Gluconeogenesis**. The liver, of course, has a few tricks up its sleeve and biochemically speaking can spin straw into gold! It can make new glucose from various odds and ends, and in particular from amino acids. This is a vital part of the response and so glucagon stimulates gluconeogenesis.

- **Fatty acid metabolism.** It is not an appropriate time to be making new fatty acids, and on the other hand the liver needs to have an energy source itself to drive all its activity. Fatty acid β-oxidation in the liver therefore needs to be up-regulated and fatty acid synthesis turned down.

These predictions, needless to say, are precisely what glucagon does. It is another example of a hormone that acts via adenyl cyclase and cyclic AMP (Topic 41).

Adrenaline

The third of our set of hormones to consider is adrenaline (or as American scientists prefer to call it, epinephrine – both names are disguised Latin or Greek ways of saying that the hormone comes from a gland beside the kidney!). We already mentioned adrenaline in Topic 41 as a prototype example that has led to much of our more general understanding about hormone action in general. Where does it fit into our developing picture with insulin and glucagon? Like glucagon, adrenaline is a mobilising hormone, but on a very different timescale. Glucagon can adjust what happens over many minutes and hours; adrenaline by contrast is termed the 'fight-or-flight' hormone and needs to produce a response to danger in seconds. Its release from the adrenal medulla is therefore triggered by impulses from the nervous system. Unlike glucagon but like insulin, adrenaline needs to be able to act all over the body. Its effects on the liver are similar to those of glucagon but more immediate and larger. Also, however, adrenaline has a major effect on adipose tissue, causing **lipolysis** (fatty acid release), thus mobilising a second energy source. Crucially, it has a dramatic effect on muscle, both skeletal and cardiac. As in liver, it mobilises glycogen (by the cAMP-dependent mechanism explained in more detail in Topic 41), but unlike the situation in liver, in muscle adrenaline greatly stimulates glycolysis. Thus, here we have a single hormone exercising opposite metabolic effects on two different tissues.

We already saw in Topic 41 how the cascade effect leads to a massive and rapid amplification of the initial signal resulting from adrenaline binding to its receptor. We now know also that the subtleties of different responses to adrenaline are in part brought about by the existence of more than one class of adrenaline receptor. There are so-called α- and β**-adrenergic receptors**, and some commonly used drugs are specifically intended to damp down the response to the hormone; thus, β **blockers** are widely used in the control of hypertension or high blood pressure.

In each of the cases we have discussed, the hormone brings about its response by the mechanisms described in the last topic, i.e. by increasing or decreasing the activity of particular enzymes. In most cases, one particular enzyme can be identified as the bottleneck, the 'rate-limiting' enzyme. Up- or down-regulating the activity of such an enzyme is in effect like turning a tap on or off.

Self-test MCQs on Topic 43

1 Which of the following statements about hormone action is true?

(a) Glucagon and adrenaline both produce short-term responses, whereas insulin produces a longer-term response.

(b) Adrenaline and insulin both produce short-term responses, whereas glucagon produces a longer-term response.

(c) Insulin produces a short-term response, whereas adrenaline and glucagon both produce a longer-term response.

(d) Adrenaline produces a short-term response, whereas glucagon and insulin both produce a longer-term response.

2 Which of the following statements about the sites of hormone production is true?

  (a) Insulin, glucagon and adrenaline are all produced by the pancreas.

  (b) Insulin, glucagon and adrenaline are all produced by the liver.

  (c) Insulin and glucagon are produced by the pancreas but adrenaline is not.

  (d) Insulin is produced by the liver but glucagon and adrenaline are not.

3 With regard to the breakdown of glycogen, which are the correct hormonal effects?

  (a) Adrenaline and glucagon switch it on; insulin switches it off.

  (b) Adrenaline and insulin switch it on; glucagon switches it off.

  (c) Glucagon switches it on; adrenaline and insulin switch it off.

  (d) Insulin switches it on; adrenaline and glucagon switch it off.

Diabetes

What is diabetes?

Patients and clinicians alike recognise and name diseases by their symptoms. The name diabetes comes from a Greek word referring to the fact that the affected patients pass large amounts of urine (polyuria). They are also therefore thirsty and drink large amounts to make up for the loss. Like most named diseases, therefore, diabetes comprises a set of clinical symptoms arising from a specific problem. In the case of diabetes, however, it was realised a long time ago that two quite different types of problem could cause the characteristic 'frequency' or polyuria, and so the broad term 'diabetes' had to be sub-divided. One problem relates to the control of water movements in the kidney by the antidiuretic hormone (ADH), vasopressin (briefly revisited in Topic 46). Faulty control here causes a relatively rare form of diabetes called diabetes insipidus. We shall not consider it further here, because, despite the apparently similar symptoms, this is a quite separate disease from **diabetes mellitus**. You are likely to hear more about vasopressin in your studies of physiology.

Diabetes mellitus is a serious and increasingly common condition in Western societies. It is a disease that can be managed, but over time it can lead to major complications including blindness, kidney disease, neuropathy and circulatory disorders that can ultimately lead to progressive amputations of necrotic appendages. But what is it?

The problem in the case of diabetes mellitus is related to the control of blood glucose levels, and, once clinical investigation of the patient's condition starts, the most immediately obvious indicator is **hyperglycaemia**, abnormally elevated blood sugar. As we saw in the last topic, the passage of sugar from the blood into the tissues is hormonally regulated, and in diabetes that regulation has failed, so that the 'on switch' is not working properly. This means that there is far too much sugar

Pain-Free Biochemistry   Paul C. Engel
© 2009 John Wiley & Sons, Ltd

in the blood (and accordingly, via the kidneys, also in the urine, allowing a simple dipstick test in the doctor's surgery). It also means, by the same token, that the tissues are being starved of glucose. This always seems like a contradiction – how can the tissues be starved when the blood glucose is so high? But the blood glucose is high precisely because the tissues *are* being starved.

As we have already mentioned, elevated blood glucose levels (above 10 mM) provide the key clinical yardstick for diabetes, along with glucose in the urine. The blood glucose also needs to be regularly measured in cases under treatment. For initial diagnosis, however, one of the most reliable indicators is the **glucose tolerance test**. This involves monitoring the blood glucose levels over time after giving a large measured oral dose of glucose. In a healthy individual the glucose level drops quite rapidly back to normal after a temporary peak, but in a diabetic the levels stays high for several hours.

Different kinds of diabetes mellitus: type 1 diabetes

So we have a definition of the disease, but, as is often the case, once we dig deeper it turns out that there are still different ways that the same symptoms and clinical indicators can arise. As explained in the last topic, there is a molecular 'gatekeeper' in the cell membranes of our muscle and adipose tissue, which controls the entry of glucose, and it is under the control of the hormone insulin. The gatekeeper is the **insulin receptor**.

If we think of these two as a key and a lock, then it is immediately obvious that if the lock cannot be opened it might be because the lock is faulty, but equally the key might be faulty or simply lost. In fact, there are two broad classifications of different types of diabetes mellitus on exactly this basis. **Type 1 diabetes**, also known as early-onset diabetes or insulin-dependent diabetes mellitus (IDDM), is the one with the missing key: there is insufficient insulin and usually this arises because of an auto-immune attack on the β-cells of the pancreas, which produce the insulin. (However, now that you have learnt about proteins and genes, you might be able to envisage ways in which the key could be defective rather then missing!)

Insulin therapy

In type 1 diabetes, the treatment is obvious and effective; the patients need more insulin and this can be self-delivered by injection. This only became a possibility after insulin was discovered in the 1920s by Banting and Best, and it raises new practical questions – where do we obtain the insulin and how do we process it? For a long time the insulin used in medical practice was from either pig or ox pancreas. Both of these work well, but there is a problem, since the amino acid sequence of these non-human hormone proteins is not 100% identical to that of human insulin. As a result, some diabetic patients developed an immune response to the foreign protein, leading first of all to allergy but also secondarily to insulin resistance. In recent years, molecular genetics has provided a way round this problem. The human gene for insulin has been cloned and can be expressed in bacteria. We do not, of course, normally raise an immune response to human insulin, provided that all the bacterial proteins have been carefully removed.

Apart from the symptoms we have already discussed, type 1 diabetics tend to present with another characteristic readily picked up by an alert clinician. Since sugar is not being properly handled, the tissues have to obtain more of their energy from fatty acid oxidation, but the capacity of the Krebs cycle tends also to be impaired, so that in uncontrolled type 1 diabetes there is a pile-up of ketone bodies made by the liver. This is discussed in more detail in Topic 21. Acetoacetic acid, one of the two major ketone bodies, spontaneously breaks down to acetone, which, being very volatile, is easily lost into the air the patient breathes out. The smell of acetone on the breath is unmistakeable. This should not of course be detectable if the diabetes is well managed.

Type 2 diabetes

The second type of diabetes, type 2 diabetes, is also known as late-onset diabetes or **non-insulin-dependent diabetes mellitus (NIDDM)**. This is causing great consternation in public health circles, not only because it accounts for 80–90% of cases of diabetes, but because it is rapidly on the rise. This is because it is strongly promoted by obesity, and the combination of overeating and increasingly sedentary lifestyles in urban societies is leading to an explosion of obesity, even among the young. Type 2 is a form of diabetes where the problem is with the lock rather than the key, i.e. it is the insulin receptor that is the problem, not the supply of insulin itself. Once again, you can envisage possible hereditary problems that could, rarely, produce a defective receptor, but that is clearly not what is happening in late-onset diabetes, both because it is so common and because the receptor is evidently functioning properly in early life. It seems that in obesity there are raised levels of two proteins, TNFα and resistin, which actually impair the function of the insulin receptor. Unlike the type 1 cases, these people have high levels of insulin, but their tissues are unable to respond. Injecting insulin is not a solution for these patients, and one of the most effective treatments in dietary management to restore normal body weight.

Self-test MCQ on Topic 44

Which of the following describes diabetes mellitus?

(a) Normal glucose transport into the bloodstream is impaired, and hence tissues are starved of glucose.

(b) Blood glucose level is elevated so that tissues receive toxic levels of glucose.

(c) Normal drain of glucose to other peripheral tissues leaves insufficient for needs of brain.

(d) Impaired glucose transport leaves tissues starved of glucose, although blood level is high.

Steroid hormones and receptors: fertility control, pregnancy testing, etc.

The steroids are a large group of related lipid compounds manufactured by various of our glands, notably the sex glands and the adrenal cortex. They are all derived from a single compound, cholesterol (Fig. 45.1). As we mentioned back in Topic 21, this compound itself is synthesised from HMG CoA (also the ketone body precursor). As you can see from the many numbered positions, this is a structure with the potential to have different chemical adornments at numerous places. In fact positions, 3, 11, 17, 18, 20, 21 can all carry substitutions such as $-OH$ CH_3 or $=O$ groups, and these can be in various combinations in different steroid compounds. There can also be extra double bonds in ring A. There is an army of enzymes to carry out the interconversions. There is absolutely no point in your trying to learn and remember the structures of this large set of molecules, unless perhaps you are going to work in an endocrinology unit. What is important to realise is that they are closely related and yet have dramatically different effects. For example, the female hormone **progesterone** (secreted by the corpus luteum, helping to maintain the endometrium ready for implantation and acting also on the mammary glands) is a precursor (three steps) of the male hormone **testosterone**, which is required for spermatogenesis and for producing male secondary sexual characteristics! Testosterone is in fact produced both by the testes in men and by the ovary in women, and can be converted in just one enzyme reaction into **oestradiol**, which once again is a female hormone (note that American texts drop the 'o' off oestradiol, oestrogen, etc. Estradiol is not a different hormone, just the same one on the other side of the Atlantic!). Both males and females produce both 'male' hormones, like testosterone,

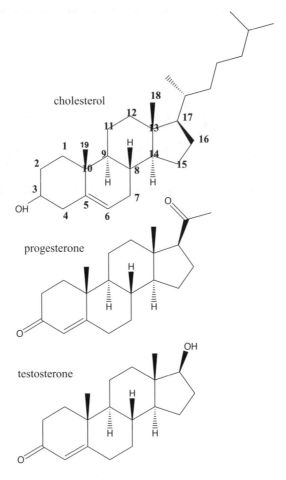

Figure 45.1 Structures of the steroid precursor cholesterol and two of the steroid hormones, showing their similar structures.

Box 12   Steroids and Olympic athletes

International competitive sport has been beset in recent years by a string of scandals relating to unfair advantage through the use of performance-enhancing drugs. The notorious examples of Eastern-bloc weightlifters, shot-putters, etc. before the fall of the Iron Curtain were among the many reasons for setting up the World Anti-Doping Agency and the establishment of procedures for routine testing of athletes. This sadly has revealed how widespread drug use has become and there seem to be constantly cases of sprinters or swimmers being stripped of their medals, Tour de France cycling teams being sent home, etc.

There are various types of performance-enhancing drugs in use, but among the most common are anabolic androgenic steroids such as testosterone, used in training to build muscle mass and increase competitive aggression. Testing procedures rely on taking closely observed urine samples of at least 75 ml, which are split into two samples and sealed in tamper-proof bottles before sending off to the analytical laboratory. The laboratory can reliably identify the dissolved substances in the urine by GC/MS analysis (see Appendix 16).

A number of problems arise. Although it is relatively easy to identify compounds that are non-physiological (e.g. synthetic steroids that are not chemically identical to our own natural hormones), it is less easy to identify drug use if the compound being used *is* the natural hormone. One method that can help is the measurement of isotope ratios (see Chemistry X). Testosterone taken as a drug is often derived from plant steroids, which have a different ratio of ^{13}C to ^{14}C and this can be readily detected by mass spectrometry. However, the athletes and their drug suppliers can get round this by providing steroid derived from animal sources. Another useful indicator for the analyst is the ratio of testosterone to epitestosterone, which through our normal metabolism would have a value below 6. Taking large doses of testosterone will, of course, elevate this ratio. This in turn leads the dopers to use masking agents; a dose of epitestosterone shortly before an anticipated test would bring the ratio apparently back to normal.

There are also potential problems relating to the handling and storage of the sample. If it is not refrigerated or frozen and if it is not treated with antibacterial agents, bacteria may metabolise steroids in the urine sample, altering the pattern that confronts the analytical lab. Another issue is that an athlete may innocently have eaten meat from poultry or other farm animals treated with steroids to enhance *their* muscle mass!

Ever greater care and scientific ingenuity is going into testing procedures and these are certainly turning up offenders with depressing regularity. However, a look at the Internet will also reveal that the test procedures have stimulated a growth industry in which equal scientific ingenuity is being applied to ways of eluding detection!

and 'female' hormones, like progesterone and oestradiol. What matters is that they are produced in the right balance at the right time.

As with other hormone effects, steroid hormones act by attaching to a protein receptor molecule. The fact that such small chemical differences among these basically very similar steroid molecules can result in such profoundly different effects is just another indication of the extraordinary power of selective molecular recognition by protein molecules. Unlike the receptors for many other hormones, receptors for steroid hormones are not embedded in the cell membrane. They do not need to be because these hormone molecules are both small and hydrophobic in nature, so that they can readily enter and cross membranes. The receptors are found in the cell nucleus, and these hormones, once bound to their specific receptors, tend to act by attaching to DNA and altering the expression levels of particular genes.

Steroid hormones, or synthetic analogues of the natural hormones can themselves be used as drugs. Mixtures of oestrogen and progestogen in various doses form the basis of (1) the contraceptive pill (stopping ovulation), (2) the emergency 'morning-after pill' (preventing implantation) and (3) hormone replacement therapy (HRT) to combat effects of ageing such as osteoporosis. An important tool in relation to such therapies is the ability to measure a woman's natural hormone levels, and here the use of specific antibody detection methods for each hormone has been of enormous benefit. Initially this was done by **radioimmunoassay**, and Rosalind Yalow won a Nobel Prize in the 1960s for introducing this methodology. Now, however, simpler, cheaper and safer methods are available, typically producing a measurable colour, but still making use of specific antibodies.

One other test that should be mentioned in this context is the **pregnancy test**, which has been streamlined to the extent that any woman can perform it at home. This, once again, is an antibody-based test, but in this case not for a steroid molecule. The test is for a protein hormone called **chorionic gonadotropin** (hCG). This hormone maintains the corpus luteum and stimulates progesterone secretion, and is produced by cells of the developing embryo and the placenta. Its concentration rises rapidly and dramatically in the early weeks of pregnancy. Since it is not produced in similar amounts at other times, it is a very reliable indicator.

Understanding of steroid action and of the role of receptors has led to some important therapeutic agents. Thus **tamoxifen**, effective against some breast cancers, is an antagonist for the binding of oestrogen to its receptor. Similarly, several drugs are available to antagonise the binding of testosterone to its receptor in prostate cancer, and the survival rate in this condition has dramatically improved in recent years.

Self-test MCQs on Topic 45

1 Which of the following statements about steroid hormones is false?

(a) Steroid hormones typically bind to a specific receptor in the cell membrane.

(b) Steroid hormones are all ultimately produced from cholesterol.

(c) Levels of steroid hormones may be measured with specific antibodies.

(d) Structurally similar steroids may produce very different physiological effects.

2 Which of the following statements about testosterone is true?

(a) Testosterone is exclusively a male hormone.

(b) Testosterone is only found in women in cases of ovarian cancer.

(c) Testosterone is made by the ovaries as an intermediate on the way to oestradiol.

(d) Testosterone is antagonised by tamoxifen in the treatment of breast cancer.

TOPIC 46

Pituitary hormones and feedback loops

The pituitary gland (also known as the hypophysis) is found at the base of the brain and together with the neighbouring hypothalamus forms a master control box for the entire endocrine system. The pituitary produces a range of 'trophic' hormones, which switch on activity elsewhere in the body. These include:

- prolactin, activating the corpus luteum and the mammary gland

- luteinising hormone (LH) acting on ovaries and testes

- follicle-stimulating hormone acting in harness with LH in both males and females

- adrenocorticotropic hormone (ACTH), stimulating the adrenal cortex;

- thyroid-stimulating hormone (TSH)

- growth hormone (somatotropin) with a profound effect throughout the body;

- melanocyte-stimulating hormone promoting skin pigmentation.

In addition to these trophic hormones, the pituitary is also responsible for the production of vasopressin (ADH) and oxytocin. Vasopressin is produced (or not) in response to the osmotic balance so that, if necessary, water reabsorption in the kidney is promoted, resulting in a more concentrated urine. Oxytocin is a hormone of pregnancy, associated both with uterine contraction in childbirth and milk ejection in breast-feeding.

Pain-Free Biochemistry   Paul C. Engel

The pituitary hormones we are discussing are all polypeptides of various lengths, ranging from growth hormone and prolactin, both about 200 amino acids long, to ACTH (39 amino acids) and even down to Met-enkephalin (5 amino acids). One of the remarkable features that has emerged is that the gene-encoded precursor tends to be a longer polypeptide, which can be cut up into several different hormone molecules by proteolytic enzymes!

The release of these pituitary hormones is itself controlled by the hypothalamus, which produces a series of releasing hormones and release-inhibiting hormones, which are selectively fed straight to the adjoining pituitary in response to neural and metabolic signals. This system provides a cascade effect: a tiny amount of hormone released from the hypothalamus leads to the release of a larger amount of hormone from the pituitary, which then reaches a target organ, say the adrenal cortex in the case of ACTH, and triggers in turn the release of a still larger amount of the characteristic product of that organ: cortisol in the specific case mentioned (see Topic 48). The other important feature of these regulatory systems is the element of **feedback control**. Once the switch is on, what is to switch it off? One answer lies in the removal of hormone molecules once they have delivered their message, but the other lies in feedback. Thus, in our example, cortisol, the end-product hormone, exercises a negative control on the hypothalamus and the pituitary, essentially informing them that there is enough cortisol in circulation, so that their positive signals are switched off. There is analogous feedback on the pituitary by thyroid, ovarian and testicular hormones. Overall, in each case the system is designed to achieve a rapid response and to return the body as soon as possible to an appropriate balance.

Self-test MCQs on Topic 46

1 Which of the following hormones is not produced by the pituitary gland?

(a) Growth hormone (somatotrophin)

(b) Glucagon

(c) Vasopressin

(d) ACTH

2 Which of the following correctly explains feedback control between pituitary and thyroid glands?

(a) Pituitary puts out a hormone that promotes thyroid activity.

(b) Thyroid puts out thyroxine, which promotes pituitary activity.

(c) Pituitary puts out a hormone that switches off thyroid stimulation of the pituitary.

(d) Thyroid puts out thyroxine, which switches off pituitary stimulation of the thyroid.

TOPIC 47

Thyroid hormones

The thyroid gland in the neck has an appetite for iodine, a chemical element that we do not encounter anywhere else in our metabolism. More accurately, it is iodide that the gland takes in (this is the negative ionic form I^- found in combination with positive ions to forms salts, such as sodium iodide (see Chemistry II). This is an example of 'active transport' (Topic 50): the thyroid cells pump in the iodide from the blood in such a way that the concentration of iodide in the gland is very much higher than anywhere else in the body.

The thyroid uses iodide for the specific purpose of making its hormone, **thyroxine**. To start with, the thyroid follicular cells make a protein called thyroglobulin. They have an enzyme that catalyses the insertion of iodine atoms into a number of amino acid (tyrosine) residues in this protein. Some tyrosine residues receive two iodine atoms, others only one. These amino acid residues then combine to form either thyroxine, with four iodines, or tri-iodothyronine, with three (Fig. 47.1). These two are also often abbreviated simply as T4 and T3, reflecting the numbers of iodines they carry. At this stage the hormone structures are still part of the protein molecule and so the final stage is that the thyroglobulin protein is processed by proteolytic enzymes to release T4 and T3 which go out into the bloodstream and round the body.

Box 13   Thyroid tumours, Chernobyl and iodide

In 1986, there was a major accident at the nuclear power plant at Chernobyl, in what was then the Soviet Union, now Ukraine. This resulted in a large release of radioactive material into the atmosphere. This was carried across Europe on the winds and resulted in heavy local pollution in neighbouring Belarus. Such

---

incidents always release ^{131}I, an iodine isotope (see Chemistry X), which has a very short half-life (8 days), which means that it is very radioactive over a short period of time. A number of other isotopes are also released, but what makes this one particularly harmful is the fact that the body concentrates iodine so effectively in the thyroid gland. In parts of Belarus, children who were below the age of 4 at the time of the Chernobyl disaster have a probability of over 30% of developing thyroid cancer. This is not an isolated incident since a number of nations through the 1950s and 1960s carried out nuclear tests in the atmosphere, again causing extensive radioactive fallout.

Although ^{131}I is a serious threat and has already caused much illness, it is also a threat that can be averted very simply. Some countries issue potassium iodide tablets (non-radioactive!) to their citizens. If these are taken in the wake of any news of a nuclear accident, then they would swamp the effect of any ingested radioactive iodide. Our thyroid gland cannot distinguish radioactive and non-radioactive iodide, and so, if, for the sake of argument the iodide tablet supplies a level of non-radioactive iodide a hundred times higher than the level of radioactive iodide taken in, this will decrease the effective risk by 99% by simple dilution.

It should also be mentioned that the thyroid gland's amazing ability to concentrate iodide is also put to use both diagnostically and therapeutically with the help of radioactive isotopes, with due thought and care. ^{123}I has an even shorter half-life than ^{131}I, so that a low dose of this isotope will be concentrated in the thyroid, allowing diagnostic imaging of the gland, but will soon decay, causing no harm to other tissues and only a brief exposure in the thyroid itself. If diagnosis reveals a tumour, then treatment with the longer lasting ^{131}I may be deliberately given. There might seem to be a contradiction here – on the one hand avoiding ^{131}I in order to minimise the risk of thyroid cancer, and on the other hand administering ^{131}I to treat thyroid cancer. This is reminiscent of the issues you will find in Topic 40 in relation to chemotherapy; rather more drastic approaches are justified at the point where it becomes necessary to kill cells.

In view of the effect of T3 and T4 on energy metabolism, individuals with an over-active thyroid tend to be hyperactive, hot and thin. Conversely, an under-active thyroid leads to sluggish behaviour and the patient tends to feel cold. The thyroid is itself under hormonal control by the pituitary via TSH, and, as a result, an under-active thyroid leads in some cases to **goitre**, with its characteristic thickening of the neck. Historically, this was common in regions where the drinking water had unusually low levels of iodide. Because the thyroid was not able to make enough thyroxine, there was no feedback signal to the pituitary to switch off TSH and so the thyroid kept growing. In order to avoid this, many countries insist that table salt has sodium iodide added to it.

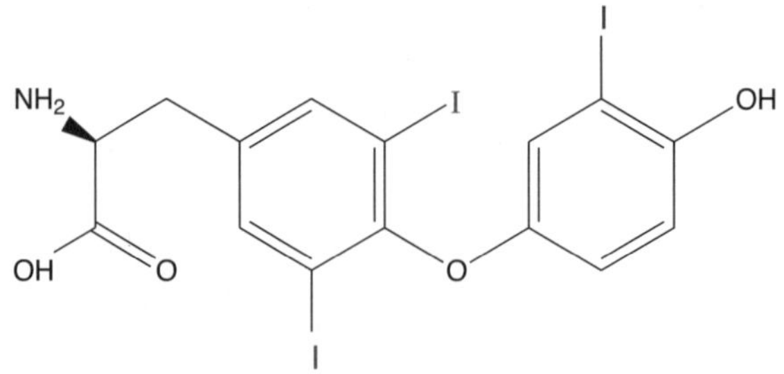

triiodothyronine (T₃)

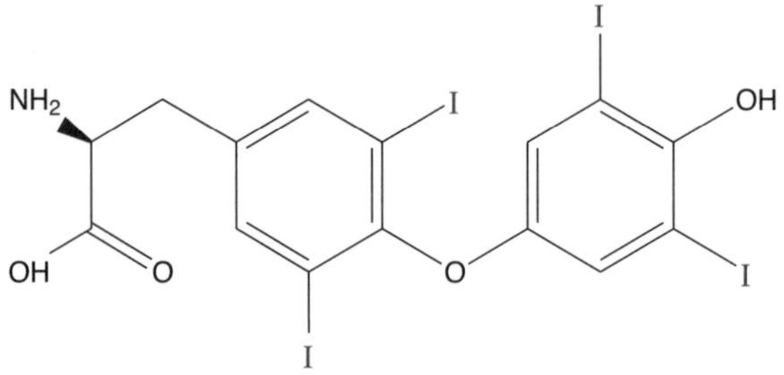

thyroxine (T₄)

Figure 47.1 Thyroxine (T4) and triiodothyronine (T3), the thyroid hormones.

Thyroid hormones act on tissues all over the body and have far-reaching effects on energy metabolism, turnover of proteins, lipids and carbohydrates and consequently growth. Thyroid deficiency in early childhood, if untreated, leads to stunted growth and poor development of the nervous system. Such individuals are termed **cretins**. This is a word that has unfortunately passed into common usage as a term of abuse but refers to a specific clinical condition brought about by endocrine insufficiency.

Self-test MCQs on Topic 47

1 Thyroxine is which one of the following?
  (a) A protein
  (b) An amino acid
  (c) A steroid
  (d) A receptor

2 Synthesis of thyroxine requires which element that is relatively unusual in a biological molecule and needs to be supplied through the diet?

(a) Fluorine

(b) Selenium

(c) Iodine

(d) Bromine

(e) Silicon

3 Excessive levels of thyroxine cause which one of the following?

(a) Hyperactivity

(b) Cretinism

(c) Hyperglycaemia

(d) Goitre

TOPIC 48

Adrenal cortex

Small though they are, the adrenal glands are really two glands in one. We have already met adrenaline, produced by the adrenal medulla. The adrenal cortex produces different hormones. One is **aldosterone**, important in ionic balance. It acts on the kidney tubules and controls the energy-dependent pumping and reabsorption of sodium ions from the urine. The mechanism for such pumping is something we shall revisit in Topic 50. The other is **cortisol**, a so-called glucocorticoid, because an important facet of its function is its action on carbohydrate metabolism. Cortisol, like the sex hormones discussed in Topic 45, is a steroid, derived from cholesterol, and as we have seen in Topic 46, its production is the culmination of a chain of hormonal events starting in the hypothalamus. Like the release of adrenaline from the adrenal medulla, the release of cortisol from the cortex is a stress response, but a longer-term response. Adrenaline (Topics 41 and 43) provides an immediate release of glucose from polysaccharide stores. Cortisol, on the other hand, mobilises depot fat and also muscle protein as alternative sources of energy and stimulates gluconeogenesis.

Cortisol (hydrocortisone) is also used in clinical practice as an **anti-inflammatory** agent, e.g. in asthma, arthritis, psoriasis, etc. There is, however, a significant risk in the regular use of such hormone therapy in chronic conditions. First of all, the mobilisation effects of the hormone lead to muscle wasting and therefore weakness, to bone demineralisation, to redistribution of fat leading to 'moon face', etc. Secondly, among the anti-inflammatory effects is a general damping down of the immune response, laying the patient open to infection. Thirdly, because of the feedback systems discussed earlier, a constant external supply of adrenal hormone provides a signal that will lead the pituitary to stop producing ACTH, thus decreasing a patient's own ability to produce the hormone normally. In view of this, it is essential to decrease the dosage gradually over time, as to discontinue it suddenly would precipitate a crisis.

Pain-Free Biochemistry   Paul C. Engel
© 2009 John Wiley & Sons, Ltd

The effects of prolonged cortisol treatment, not surprisingly, are similar to the symptoms of Cushing's syndrome, a long-recognised condition resulting from an over-active adrenal cortex.

Self-test MCQs on Topic 48

1 Identify the false statements among those below.

   (a) The adrenal cortex produces cortisol and adrenaline.

   (b) Cortisol is a steroid.

   (c) Cortisol is a glucocorticoid.

   (d) Cortisol promotes the deposition of fat and muscle protein.

   (e) Cortisol counteracts inflammation.

   (f) Cortisol promotes bone deposition.

2 The term 'adrenal cortex' refers to which one of the following?

   (a) A metabolic crisis when steroid therapy is withdrawn too rapidly from a patient.

   (b) A distinct region of the adrenal gland producing aldosterone and cortisol but not adrenaline.

   (c) The complex of hormones produced by the adrenal gland.

   (d) A congenital condition in which an over-active adrenal gland produces growth abnormalities.

TOPIC 49

Prostaglandins and inflammation: aspirin

Earlier in the book, in Chemistry IX and Topic 22, we came across the idea of unsaturated fatty acids, i.e. fatty acids with double bonds in their long chains of linked carbon atoms. We saw that some of the polyunsaturated fatty acids, with three, four or more double bonds, cannot be made by our own cells and so have to be taken in through our diet – essential fatty acids. So far, however, we have not seen a role for these fatty acids.

One of them is linolenic acid, an 18-carbon fatty acid with 3 double bonds. Our metabolism can convert this essential fatty acid into **arachidonic acid**, which is a 20-carbon fatty acid with 4 double bonds. It is named after the peanut (*Arachis hypogaea*), since it is found in peanut oil, and so we can also obtain it directly in the diet. We definitely do know what the role of arachidonic acid is because it is the precursor to a class of compounds called eicosanoids (from the Greek word for 20, because they all have 20 carbon atoms). These include two classes of potent compounds called **prostaglandins** and **thromboxanes**. The name of the first class indicates an association with the prostate gland early in the discovery and study of these compounds, but it is now realised that both classes are very widespread in the body and are important in cell signalling processes. Specifically they are intimately involved in inflammation and pain. We tend to view both inflammation and pain as bad things, but they are both there for a purpose. Nevertheless, they can get out of hand, and it is an important aim of pharmacology to bring them under control.

There are many examples of traditional remedies that have found a biochemical explanation only in retrospect, although nowadays we try to apply our biochemical knowledge to intelligent, rational design of new drugs. One of the most striking

Pain-Free Biochemistry   Paul C. Engel
© 2009 John Wiley & Sons, Ltd

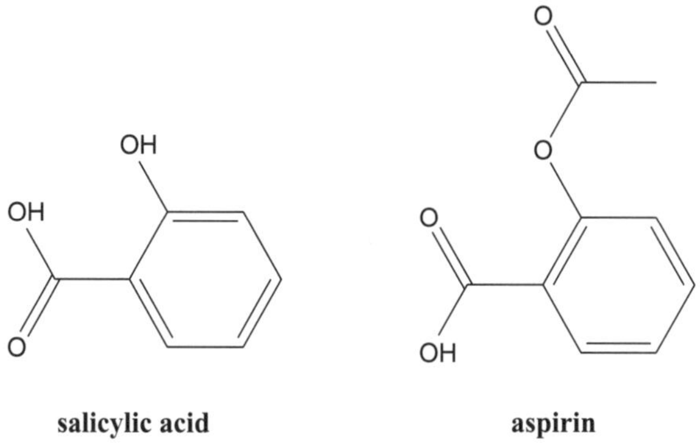

salicylic acid   aspirin

Figure 49.1 Salicylic acid and aspirin.

examples of the first situation is **aspirin**, which must be one of the most widely used of all drugs. The traditional remedy was an extract of willow bark to control fever and pain, and the active ingredient, salicin, is converted in our bodies into salicylic acid (Fig. 49.1). Aspirin is acetyl salicylic acid, a synthetic derivative, which is pleasanter to take, being less bitter, and its appearance on the market over a hundred years ago was the first big success for a new industry, the pharmaceutical industry.

Today we know exactly how aspirin works. Arachidonic acid is the substrate for an enzyme called **cyclooxygenase**, often abbreviated as COX (and now COX-1 and COX-2, because we all have two different versions with different properties). Cyclooxygenase activity converts arachidonic acid to an essential precursor on the road to making all the eicosanoids. What aspirin does is attach itself in the active site of cyclooxygenase, and it permanently inactivates the enzyme by chemically transferring the acetyl (CH_3CO-) group from the drug molecule to the enzyme molecule and so irreversibly blocking the active site (Appendix 14). The fact that this is irreversible means that dosage is rather important with a drug like this. The enzyme is there for a purpose after all, and totally wiping it out would not be a good idea. Damping it down, however, has proved to be a very useful way of controlling pain. Two more very widely used drugs have followed, targeting the same enzyme, acetaminophen or **paracetamol**, and **ibuprofen** (Fig. 49.2).

The compounds mentioned here are examples of so-called non-steroidal anti-inflammatory drugs (**NSAIDs**). Pharmacologists are always on the lookout for greater selectivity and fewer side effects in drug action, and steroids, while being very effective anti-inflammatory drugs, also have a number of much less desirable side effects, e.g. demineralisation of bone, hence the interest in developing new NSAIDS. As mentioned above, it is now known that there are two different COX enzymes, and one of the current targets in drug development is for greater selectivity in the **inhibition** of these two related enzymes.

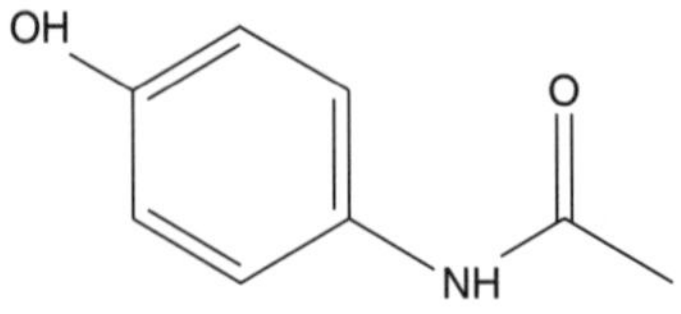

paracetamol

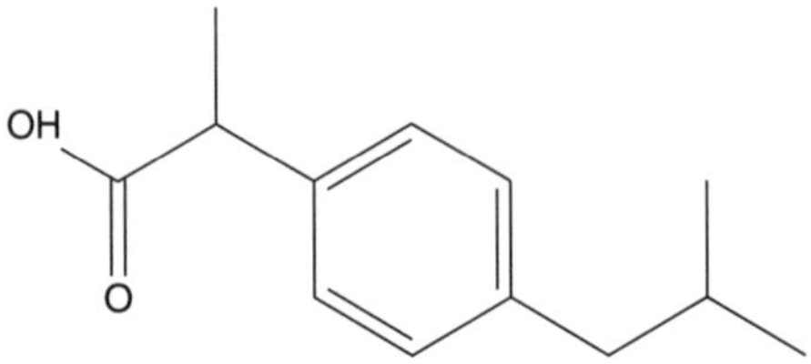

ibuprofen

Figure 49.2 Paracetamol and ibuprofen.

Self-test MCQs on Topic 49

1 Which one of the following is true?

(a) Arachidonic acid is important in our diet because it can supply linolenic acid for prostaglandin formation.

(b) Prostaglandin is important in our diet because it can supply linolenic acid for arachidonic acid formation.

(c) Linolenic acid is important in our diet because it can supply arachidonic acid for prostaglandin formation.

(d) Arachidonic acid is important in our diet because it can supply prostaglandin for linolenic acid formation.

2 Linolenic acid is which one of the following?

(a) A polyunsaturated fatty acid.

(b) A steroid.

(c) A polysaccharide.

(d) A thromboxane.

3 Aspirin works by doing which of the following?

(a) Blocking the absorption of linolenic acid.

(b) Blocking pain receptors.

(c) Blocking the synthesis of salicylic acid.

(d) Blocking the enzymic oxidation of arachidonic acid.

4 Ibuprofen is which one of the following?

(a) An non-steroidal anti-inflammatory drug.

(b) A thromboxane precursor.

(c) A component of fish oils.

(d) Another name for aspirin.

Membrane transport

The need for communication between compartments

As we have seen elsewhere, the cell is divided into various compartments, nucleus, mitochondria, cytosol and so on, each one separated from the rest by membrane barriers. If everything could move freely across these membranes there would be no point in having them. They would be like national borders with no one checking passports, visas, smuggling, etc. The organelles keep certain processes inside their boundaries, e.g. the Krebs cycle, which takes place exclusively inside the mitochondria, or anaerobic glycolysis exclusively in the cytosol, etc. On the other hand, if there were no communication and traffic between compartments, there would in effect be no cell and no organism – the whole has to be more than the mere sum of its parts, and these parts are not set up to be autonomous but to carry out specialised functions as part of an overall team effort.

Passive transport

The simplest aid to transport is a 'pore' or channel, i.e. a hole across the membrane. A hole in a biological membrane, however, cannot just be a hole. First of all, the nature of the lipids that make up our membranes is such that any such hole would instantly seal itself up again. Also however, a simple hole would mean that anything and everything could leak out, in which case it would be pointless to have the membrane barrier at all. So, for there to be a permanent pore or channel, there must be a specific protein that sits in and across the membrane to provide the perimeter of the hole. The fact that such a hole will have a precise size and that it will be lined with amino acid side-chains means in fact that it will be selective. Unlike a hole in, say, a fence, it will not automatically let anything pass that is small enough, but only those things that are of the right size and characteristics. The proteins responsible for some of these gateways in the membrane are called '**porins**', and they include the aquaporins, which very specifically allow water molecules in and out, especially in places like the kidney tubules where there is a need for rapid movement of water across the hydrophobic lipid membrane.

There are also various selective **ion channels**, for K^+, for Na^+ and for Ca^{2+}, for example, and selective channels for such key compounds as glucose. The **glucose transporter** is under the control of insulin as we saw in Topic 43. These, however, are passive transporters, i.e. no energy is required for their action; they just let substances in or out, moving in the direction dictated by the concentration gradient, i.e. from more concentrated to less concentrated. There are, however, also a number of locations and situations that demand actual pumping uphill against a concentration gradient. This is dealt with in the next section.

Ionic gradients, Na^+/K^+ balance

If one measures the concentration of ions inside and outside our cells, there is normally an overall balance so that there is not an osmotic gradient across the membrane. However, when one looks more closely at which ions are where, it turns out that there are striking gradients of individual ions. Among the most dramatic are the relative levels of Na^+ and K^+ ions (Fig. 50.1). Chemically sodium and potassium ions are thought of as very similar, but living systems have chosen to treat them as critically different. Na^+ in the extracellular fluid is present at about 140 mM, but inside the cells its concentration is only 10 mM, 14 times lower! Conversely, K^+ is present at only 4 mM outside the cells but its intracellular concentration is 140 mM, presenting an even steeper 35-fold concentration gradient across the membrane. Any passive diffusion would inevitably tend to eliminate these gradients and the fact that they are there means that there must be an active transport process maintaining them. It is in fact a single active transport pump, the **Na^+/K^+ ATPase**, that is responsible for both gradients. The name of this pump tells us that it requires both

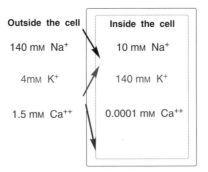

Figure 50.1 Cation gradients across the cell membrane.

types of ion in order to split ATP, but that emphasis reflects the way it was studied rather than what it is there for. Physiologically, it uses the splitting of ATP to drive the pumping of Na^+ and K^+ ions rather than using the ions to promote the splitting of ATP! Each cycle of its transmembrane activity splits one ATP molecule to ADP and in the process exchanges two K^+ ions going in for three Na^+ ions going out. This means that it is an electrogenic pump because its action is tending to produce a difference in charge across the membrane, a so-called **membrane potential**. Animal cells do indeed maintain a membrane potential across their cell membranes and it can be expressed just like the potential between the terminals of a battery – in millivolts. Typically the membrane potential will be about 70 mV.

These ionic gradients and the membrane potential are in a way like the drawn string of a bow or the cocked trigger of a gun, and we shall see in Topic 51 that they can be used just like a bowstring or a trigger to produce a sudden response.

Calcium ion gradients

The extracellular concentration of Ca^{2+} is approximately 1.5 mM. The cytosolic concentration, on the other hand, is only 0.1 μM (note the units; see Chemistry III). This is a gradient of more than 10 000:1 in concentration (Fig. 50.1) and again requires active transport, in this case by a **Ca^{2+}-ATPase**. This pumps calcium ions out of the cell across the plasma membrane. Alternatively Ca^{2+} can also be pumped out of the cytosol into an internal compartment, the endoplasmic reticulum. In the case of muscle, the highly specialised internal compartment is known as the sarcoplasmic reticulum and plays an important role in triggering contraction.

Symport and antiport systems

In addition to the primary active transport systems we have just considered, there are a considerable number of other specific transport systems to carry individual metabolites, ions, vitamin substances, etc. across membranes, which essentially piggyback on these primary systems. They operate on the basis either of exchanging, say, with a sodium ion, one in for the other out (this is called 'antiport') or travelling together across the membrane (this is 'symport'). In either case, the energy input comes from the original ATP-dependent pumping; the pumped ion can then run back down its concentration gradient, just like a wind-up spring unwinding, and in doing so it carries another substance across the membrane.

Transport and drug resistance

Finally, we must note that transport systems are not a unique property of our own bodies. They are, of course, an essential part of the workings of all living things, including those that find it convenient to infect us. In an earlier topic, we discussed bacterial drug resistance and the fact that many bacteria destroy antibiotics by enzymatic action. However, a growing threat comes from **multiple drug resistance**, and this frequently is due not to multiple separate enzymes each breaking down a different drug but rather to a powerful ATP-dependent pump, which is able to expel a variety of different drug molecules.

Self-test MCQs on Topic 50

1 Which of the following is true about membranes surrounding cells and organelles?
   (a) They are impermeable and so are able to keep each metabolic compartment entirely segregated from the rest.
   (b) They operate like sieves with holes that allow the passage of everything below a certain size.
   (c) They keep the physical structures of the cell separate but allow free passage of dissolved molecules so that metabolism can proceed unhindered.
   (d) They have pores that only allow the passage of specific compounds.

2 A passive transporter is which one of the following?
   (a) A transporter that only works when concentrations of the transported substance are equal on both sides of the membrane.
   (b) A transporter that only works when ATP is provided.
   (c) A transporter that only works when there is a gradient across the membrane of the substance to be transported.
   (d) A transporter that only works when it receives a specific neural or hormonal stimulus.

3 There is a gradient of sodium ions across our cell membranes. Which of the following is true?
   (a) Na^+ concentration is 3-fold higher inside the cell than outside.
   (b) Na^+ concentration is 3-fold lower inside the cell than outside.
   (c) Na^+ concentration is 14-fold higher inside the cell than outside.
   (d) Na^+ concentration is 14-fold lower inside the cell than outside.
   (e) Na^+ concentration is 300-fold higher inside the cell than outside.
   (f) Na^+ concentration is 300-fold higher inside the cell than outside.

4 Calcium ion concentrations inside and outside the cell are, respectively, which of the following?

(a) About 0.1 μM and 1.5 mM

(b) About 10 and 1.5 mM

(c) About 1.5 mM and 0.1 μM

(d) About 1.5 and 10 mM

5 Multiple drug resistance sometimes arises for which of the following reasons?

(a) The patient's cells lose the ability to transport the drugs inwards.

(b) The patient's cells acquire an ability to transport the drugs outwards.

(c) The pathogenic organism loses the ability to transport the drugs inwards.

(d) The pathogenic organism acquires an ability to transport the drugs outwards.

TOPIC 51

Nerve and muscle

Relatively large animals, such as we are, need to have a rapid system for transferring information both inwards to the brain from sense organs – eyes, nose, ears, tongue, touch sensors all over the body, balance sensors, etc. – and also back outwards from the brain to produce a rapid and appropriate response. Some reflex responses will be mediated without going all the way to the central processor in the brain, but still will involve messages being sent considerable distances. The information transfer system is, of course, the nervous system and the long axons projecting out of nerve cells are like telephone wires. But how do they work?

A stimulating signal causes a small decrease in membrane potential discussed in Topic 50, a so-called **depolarisation**. In the membrane there are sodium channels that are sensitive to this. In the normal resting state they are closed, but the drop in membrane potential to about $-40\,\text{mV}$ opens them, creating a sudden leak. As mentioned earlier, these channels are not just like a hole that will let anything through; they are highly specific, in this case for Na^+ ions. Accordingly, driven by the 14-fold gradient across the membrane, Na^+ ions flood into the nerve cell through the channels, producing a rapid reversal of the membrane potential, which goes from negative to positive ($\sim +30\,\text{mV}$) in the space of about a millisecond. This sharp change in potential now (1) closes the sodium channels and (2) opens potassium channels so that K^+ ions can stream out, tending to wipe out the transient positive potential, and with the assistance of the membrane 'ATPase' the original situation is rapidly restored (repolarisation). This process of successive depolarisation and repolarisation spreads progressively down the axon (the long wire-like projection from the main cell body of a neuron).

This, however, only tells us how to propagate a nerve impulse along a single nerve cell. What happens when the signal reaches the end of the nerve cell? The answer to this depends on what the job of the nerve cell is. If it has to pass on its

Pain-Free Biochemistry   Paul C. Engel
© 2009 John Wiley & Sons, Ltd

message to another nerve cell, then there is a **synapse**, where the surfaces of the two nerve cells come into intimate contact. The first cell passes the message across the narrow gap of the synapse by releasing **neurotransmitter** molecules. There are different neurotransmitters used at different types of synapse, including acetyl choline, noradrenaline, adrenaline, dopamine and glutamate, but the role of the neurotransmitter is to diffuse rapidly across the gap and trigger a new depolarisation so that the nerve impulse can be propagated further. (It is a failure of the neurotransmitter response that is the problem in **Parkinsonism**.)

This electrical pulse process, however, only occurs periodically along the length of the nerve fibre. If we think of the nerve axon as being like an electric cable, then, like a cable, it is well wrapped round with insulation. In the case of a nerve, this insulation is provided by the **myelin sheath** (which erodes in **multiple sclerosis**). There are gaps in the insulation every so often, and it is at these points, nodes, that the electric pulse is generated. The electric signal at each of these nodes leads to the depolarisation at the next node so that the signal is propagated along the nerve in jumps ('saltatory conduction').

Nerves, however, do not always transmit to other nerves. If you touch a hot stove, the end result of all the message transmission has to be that you pull your hand away automatically, and so in this case the final destination of the message is to a different kind of excitable tissue, muscle. The message comes down to the **neuromuscular junction**, and here the transmitter substance is **acetyl choline**. In the nerve ending the release of acetyl choline from storage vesicles is triggered by another ion channel event. The depolarisation opens up calcium channels, and the influx of Ca^{2+} ions leads to the release of the acetyl choline into the synapse.

The effect of acetyl choline in turn is to trigger muscle contraction, but before we consider that, we have to think about the fate of this chemical messenger. If it were to stay there, the muscle would keep on contracting, possibly long after the need had passed! Therefore, there has to be a way of ensuring that its effect is only fleeting. This is provided by an enzyme, **acetylcholinesterase**, which degrades the transmitter to acetate and choline. This enzyme is the target for a remarkable number of potent pharmacological agents, ranging from the American-Indian arrow poison curare to various nerve gases for biological warfare. In more peaceful applications there are various insecticides based on inhibition of acetylcholinesterase.

Inside the muscle cell there are a number of response elements helping to control a complex response, and an important part of the machinery is the sarcoplasmic reticulum, which stores up Ca^{2+} ions (see previous topic) and then releases them as part of the trigger process. Central to the whole process, however, is the activity of two proteins, **actin** and **myosin**. These are fibrous, linear proteins, and they are really specialised enzymes, which split ATP and use the chemical energy to drive, in this case, a physical process, a mechanical movement of one molecule along the other, a notch at a time.

Self-test MCQs on Topic 51

1 A nerve impulse is transmitted down a nerve cell by which one of the following?

(a) Neurotransmitter molecules travelling along the axon.

(b) Alternate movements of sodium ions and potassium ions across the cell membrane.

(c) A gradient of calcium ions down the length of the cell.

(d) A current of protons along the axon.

2 A nerve impulse is transmitted across a synapse by which one of the following?

(a) Neurotransmitter molecules travelling across the synapse.

(b) An electric discharge across the synapse.

(c) A current of calcium ions across the synapse.

(d) Opposing movements of sodium and potassium ions across the synapse.

3 The myelin sheath serves one of the following functions. Which?

(a) It provides complete electrical insulation round the nerve fibre so that the signal is not lost.

(b) It allows selective movement of the sodium and potassium ions across the sheath.

(c) It contains receptors for neurotransmitters so that nerve cells can communicate with one another.

(d) It has periodic gaps allowing a leak so that the signal can be transmitted from one gap to the next.

4 At the neuromuscular junction the incoming signal is switched off so that a single signal does not cause permanent or continuous contraction. How is this done?

(a) It relies on the instability of the acetyl choline molecule.

(b) Enzyme action.

(c) Interaction between acetyl choline and calcium ions.

(d) A second incoming signal a short time behind the first.

TOPIC 52

pH homeostasis

We have seen, while going through the book, that protein molecules carry out all sorts of functions, working as enzymes, antibodies, hormone receptors, pumps, pores, contractile molecules, and so on, and that all these functions depend on the protein's specific shape and its ability to recognise other molecules. Back in some of the earliest topics, we saw how proteins are built up of amino acids, some of them with charged side-chains, positive or negative. We know that these charges are important both for helping to specify and maintain the overall shape of the protein and also for providing specific binding sites for other charged molecules, negative sites to attract and anchor positive groups and vice versa. The charge state of all these amino acid side-chains will depend on the pH of their surroundings (Chemistry III and Appendix 1), and we might predict, therefore, that biological processes that depend on proteins (or other charged macromolecules like nucleic acids) would be rather sensitive to pH. In fact, we know that this is true from hundreds of detailed studies of the behaviour of individual proteins that have been isolated and studied on their own. In the case of enzymes, we have the concept of the pH optimum, a pH value at which the enzyme performs at its best, and sometimes the performance drops off quite steeply on either side of this optimum value.

If you think about this, it becomes obvious that any major changes in the cellular pH value could seriously upset everything. Most of the proteins are tuned to operate well just on the alkaline side of neutrality, i.e. at pH values of 7.35–7.45. (Exceptions are those proteins that are designed to work in a special acidic niche, like pepsin in the stomach or the various hydrolytic enzymes in the cell's lysosomes; Topic 23.) But how can the body maintain a constant pH value when it is constantly carrying out activities that are tending to change the pH? Anaerobic metabolism produces lactic acid. Ketogenesis produces acetoacetic and β-hydroxybutyric acids. Above all, however, cellular aerobic respiration produces CO_2, which dissolves in and combines with water to form carbonic acid, H_2CO_3:

$$H_2O + CO_2 \leftrightarrow H_2CO_3 \leftrightarrow H^+ + HCO_3^-$$

This is not a very strong acid, but it is an acid nonetheless, ionising as shown above to form the **bicarbonate** ion HCO_3^- and H^+. The control of this equilibrium is very important in controlling pH in the blood. A first point to make in this regard is that the dissolved bicarbonate offers very significant buffering capacity. What this means is that, over a pH range in which the ionisation equilibrium of H_2CO_3 is poised, the bicarbonate ion is able to soak up most of the H^+ ions as they appear, preventing or postponing a drop in pH. Some of the proteins that are present at relatively high concentrations also contribute to this buffering effect. Because both the equilibria above are reversible, we also have a ready way of counteracting a drop in pH: the blood is constantly delivering blood back to the lungs, and the lungs will, of course get rid of excess gaseous CO_2 simply by breathing it out. You can artificially accelerate this process by hyperventilating (breathing unnaturally fast) and if you do this for a minute or two you will feel quite strange; what you will have done is rapidly raised the blood pH. Under normal circumstances, we breathe, without thinking about it, at a rate governed by internal receptors that sense the levels of O_2 and pH and automatically adjust accordingly. The equilibrium between CO_2 and H_2CO_3 occurs spontaneously, but it is too important to leave to relatively slow spontaneous chemical reaction, and we have an enzyme, one of the most active of all our enzymes, called **carbonic anhydrase**, that speeds up this process, ensuring that we can form gaseous carbon dioxide fast enough to breathe it out efficiently.

A second organ that is important in our overall pH control is the kidney. We cannot get rid of lactate or acetoacetate through our lungs, but we can filter it out through the kidneys. On the other hand, the kidney can also reabsorb ions that have been filtered out and, depending on the need either to raise or to lower the pH, the kidney tubules may or may not reabsorb filtered bicarbonate. The bicarbonate is allowed to escape in the urine if the blood pH is high (too alkaline), and this helps to bring it back down.

Self-test MCQs on Topic 52

1 Carbonic anhydrase is which one of the following?

(a) A compound formed when carbon dioxide reacts with water.

(b) A protein that acts as a buffering substance by absorbing H^+ ions.

(c) An enzyme that speeds up the breakdown of carbonic acid.

(d) An acid excreted by the kidney for pH control.

2 Our bodies cannot allow the pH to vary by more than a fraction of a pH unit because otherwise

(a) proteins would be broken down

(b) membranes would become permeable

(c) there would be painful indigestion

(d) enzymes and other proteins would cease to function properly.

3 The main organs involved in pH control are which if the following?

(a) The lungs and the liver.

(b) The skin and the kidney.

(c) The skin and the intestines.

(d) The lungs and the kidney.

(e) The liver and the intestines.

Diagnostic markers: biochemical tests

The role of biochemical tests

Much of this book has been an explanation of how the normal body/tissue/cell works in biochemical terms. However, by definition, healthcare professionals are going to be dealing with the abnormal state. It is important to think, therefore, how things might change biochemically in the abnormal state, how that might help clinicians to detect the abnormal state, early if possible, and how it might help to monitor recovery. Being sent for a battery of tests is part of the routine patient experience, even if only for a check-up. Some tests will be microbiological, but very many are biochemical tests, and so we should consider what kind of thing the hospital biochemist will be looking at and looking for and also how.

What kind of sample?

The choice of clinical sample is a balance between information and discomfort or inconvenience for the patient. Providing a urine sample or a sputum sample is painless and easy. For many purposes, however, a blood sample may be more informative, requiring a trained person, care and a patient willing to face the needle! Yet more challenging is lumbar puncture for cerebrospinal fluid. For some purposes it may be important to examine or use whole tissue or cells and then biopsy may be necessary. This might be a needle biopsy for, say, liver, muscle or breast tissue, or a skin biopsy, which might be taken in order to culture fibroblasts.

Pain-Free Biochemistry   Paul C. Engel

What to look for?

Tests might be designed to discover:

1 Malfunction of an organ or system or perhaps an enzyme. This might result in too much or too little of a normal metabolite (e.g. glucose and cholesterol), a hormone (e.g. oestrogen and thyroxine) or a mineral salt (e.g. iron and calcium), or it might result in the appearance of unusual substances. Genetic deficiencies of enzymes introduce blocks in metabolic pathways, like a dam in a stream, and so frequently produce an overspill of substances into enzyme reactions in which they would not normally be involved, producing telltale metabolic markers.

2 Tissue damage. Heart attacks, tumours, cirrhosis of the liver or hepatitis, kidney failure, stroke, etc. all damage the cells of specific tissues. As already mentioned in Box 8 (Topic 26), this may lead to very obvious elevation of characteristic proteins in blood, CSF or urine.

3 Infection. Bacteria, viruses, fungi, worms, protozoan parasites, etc. all present distinctive molecular signatures and the clinical laboratory may test either directly for signs of the infectious agent or on other occasions for signs of our antibody response as an indication of recent or past exposure.

4 Poisoning. An emergency ward might be confronted with an unconscious patient after, say, an overdose, and urgently need identification of the toxic agent.

What types of tests are available?

For small metabolite molecules like glucose or cholesterol or phenylalanine there are highly specific quantitative tests that rely on using purified enzymes as tools. These tests usually involve formation of a coloured product and may be formatted for large-scale instrumental laboratory analysis, for dip-strip analysis in a doctor's surgery or perhaps for self-testing by the patient. Some enzyme-based tests have also been built into electronic sensor devices. An alternative is offered by chromatography (see Appendix 16), frequently now coupled to mass spectrometry. These methods allow a blood or urine sample to be simultaneously screened for a large number of different substances. For slightly larger biological molecules, such as hormones, as has already been mentioned in Topic 45, there is often the option of an antibody-based test. Such tests have cunningly been made incredibly sensitive by chemically attaching an enzyme molecule (e.g. horseradish peroxidase) to each antibody molecule. This means that each molecule to be detected will be able to announce its presence through the production of millions of molecules of enzyme product, typically strongly coloured. This, in effect, turns the announcement from a whisper into a shout! These tests are called **ELISAs**, standing for enzyme-linked immunosorbent assay.

Turning to proteins, there is first of all the issue of finding protein where protein should not be. Protein in the urine is a clear indication of kidney damage because the kidney tubules should filter out small molecules and retain large ones. In this case, it is not important to identify which proteins are coming through, and a non-specific generic test for protein is sufficient. However, in many other situations the valuable diagnostic information comes from identifying the actual protein. This can be done in various ways. As we saw back in Topics 3 and 4, each protein has its own individual size and shape and electrostatic charge. These three properties make it possible to separate the hundreds of different proteins in a biological fluid or tissue extract (e.g. Appendix 15). This will give rise to a typical pattern, possibly quite complex, which is recognisable to the trained eye as the normal pattern. That pattern is disturbed if either there is far too much of a normal protein or else a protein that should not be there appears.

Identification of proteins based on physical properties is simple and cheap and can be surprisingly effective, but it does not make use of the most discriminating properties of a protein, i.e. its biological properties. In the case of an enzyme, for instance, the most sensitive and unequivocal identification comes from measuring its activity. Alternatively, a method that can equally be used for proteins that do not have catalytic activity is immunological detection. There are commercially available antibodies that have been raised against most clinically interesting proteins, and this also allows for very sensitive and certain detection. Immunological detection with specific antibodies is also the method of choice for identifying many kinds of infectious agent.

Toxicology, on the other hand, is more likely to based on chemical methods of separation and identification such as chromatography and mass spectrometry (Appendix 16).

Self-test MCQs on Topic 53

1 A diagnostic test for the presence of a particular enzyme in a patient sample would most likely be based on which one of the following?

(a) Measuring the enzyme-catalysed reaction.

(b) Using a specific antibody for the enzyme protein.

(c) Radioactive labelling.

(d) Separation according to electrical charge.

2 A diagnostic test for the presence of a particular hormone in a patient sample would most likely be based on which one of the following?

(a) Using a specific antibody for the hormone.

(b) Using an animal-based test to measure the physiological effect of the hormone.

(c) A measurement using the specific receptor for the hormone.

(d) Crystallographic identification of the molecule.

3 Finding liver proteins in a serum sample would suggest which one of the following?

(a) Recent consumption of liver in the diet.

(b) High metabolic activity in the liver.

(c) The liver supplying proteins to other tissues in the body.

(d) Liver damage.

Blood, bleeding and clotting

Our network of blood vessels, arteries, capillaries and veins has to carry out many functions. Among these are transport functions, carrying foodstuffs round the body and carrying oxygen from the lungs to the peripheral tissues and CO_2 back to the lungs. This all has to be done with speed and efficiency, and therefore the blood is forced through the network under pressure, under which it has to pass through the tiniest of capillaries. In some parts of the body, perhaps most obviously in the digestive tract, the circulatory system also has to stand up to considerable physical buffeting. It is constantly on the cards that a small blood vessel will give way under the strain, and that under the pressure of the cardiac pump the blood will burst through.

Such a **haemorrhage**, if it continues, poses a very serious threat. First of all sheer loss of blood and blood pressure is a danger, but also, depending on where the bleeding occurs, the accumulation of blood in inappropriate places can be an even more rapid threat, the most obvious example being haemorrhagic **stroke** as a result of bleeding in the brain. Therefore, our circulatory system has evolved a mechanism of instant response designed to plug up the hole wherever it occurs, stopping the bleeding by formation of a blood clot.

Starting with the **clotting response**, the system's ability to swing into a full-blown response from nothing in a very short time depends on a principle we have already met in the context of hormone action, namely a cascade mechanism. In considering the adrenaline response, we saw how a series of enzymes acting each one on the next could produce a rapid amplification of an initial signal. In that case the system worked by phosphorylation, each enzyme in the sequence catalysing addition of phosphate to the protein structure of the next enzyme, switching it on. The blood clotting cascade also relies on each enzyme switching on the next but in this case the enzymes in the sequence are all proteolytic enzymes and each one

Pain-Free Biochemistry   Paul C. Engel
© 2009 John Wiley & Sons, Ltd

catalyses a proteolytic cut in the molecules of the next one in the sequence to switch it on.

Key players in the clotting process are the blood **platelets**. These little cells respond to injury by clumping and in effect forming a physical plug. A second key component is the protein fibrin. A precursor called **fibrinogen** is one of the major protein components of the blood serum, and it remains soluble and inert until it is activated through the proteolytic action of an enzyme called **thrombin**. Thrombin likewise exists in the blood as an inactive precursor, prothrombin, which only becomes active through the proteolytic action of Factor X (Roman 10). We can work on backwards through a series of similar 'Factors' each with a Roman number and present in the blood in smaller and smaller concentrations. It is not important to learn all the components of this complex sequence. It is important, however, to realise that there are in fact two converging sets of proteolytic activity leading to the activation of Factor X. These are the so-called intrinsic and extrinsic pathways. The intrinsic pathway accounts for the fact that blood will clot apparently spontaneously even in a clean glass vessel. The extrinsic pathway relies on a factor that becomes exposed when blood vessels are broken.

On the other hand, this defence mechanism has to be kept on the tightest leash possible; if the blood starts to clot when there is no haemorrhage, then there is the risk of a blockage, and such blockages, when occasionally they occur, are extremely dangerous; a **thrombosis** blocking the supply of oxygen and nutrients to the heart can be fatal, and in the brain a stroke may result not only from uncontrolled bleeding but also from a clot. Equally, in the event of bleeding followed by a clotting response, it is vital that the clotting ceases as soon as it has done its emergency job. We do not want an escalation that clots our entire circulation.

Unlike the protein kinase system we met in Topic 41, activation of the blood clotting enzymes is not reversible – the cut pieces of the protein cannot be stuck together again. However, the blood contains other proteins whose job is specifically to attach to proteolytic enzymes and inhibit them, and so these enzymes are unlikely to get far from the site of injury and activation before they are switched off by this mechanism. This explains why our entire circulation does not stiffen up as one giant fibrin clot. One further issue remains, however: in due course the fibrin clot needs to be removed as the body sets about the longer-term healing process. For this we have a specific enzyme system for fibrinolysis, i.e. to dissolve fibrin clots. It uses another very specific proteolytic enzyme, **plasmin**. Like the others we have met, this one also exists as an inactive precursor, plasminogen. It is switched on by the action of TPA, **tissue plasminogen activator**. In recent times this has become an important drug because, now that it can be produced outside of the human body by molecular genetic techniques, it is available in pure form to administer in cases of either coronary or cerebral thrombosis.

Inappropriate blood clotting can be very dangerous, as mentioned above. In Box 14 we examine one form of intervention, with warfarin, to damp down the clotting response, but that is relatively drastic in view of the potential risks of overdose, and a more routine and less severe intervention is a daily dose of aspirin, often prescribed for men beyond middle age. As explained in Topic 49, aspirin is a COX inhibitor and in consequence it inhibits the production and release of thromboxane A2 in platelets. This messenger compound is a key factor in promoting platelet aggregation, so that blocking its release nips a potential clotting event in the bud. A recent context in which aspirin is now often prescribed is long distance air travel, with its attendant risk of **deep vein thrombosis** brought on by long hours of sitting.

A major factor in routine hospital practice is the use of blood for **transfusion**. The collection and banking of this blood clearly have to involve precautions against clotting, but equally those precautions must not mean that the use of transfused blood in surgery leads to uncontrollable bleeding! Two of the main means of inhibiting clotting in blood withdrawn from the circulation are the use of **heparin** (a natural anti-clotting agent) and **citrate** (which acts by trapping Ca^{2+} ions).

The complexity of the clotting process means that there are a number of places where it can go wrong, and there are accordingly a variety of hereditary bleeding disorders that collectively come under the label **haemophilia**. Affected individuals bruise very easily, and what other people would treat as a trivial cut or scratch may lead to life-threatening, uncontrollable bleeding. Fortunately, an improved understanding of all the various clotting factors now means that (1) it is possible to identify which factor is missing and (2) national blood banks process a considerable fraction of their blood in order to obtain purified preparations of these factors, which can then be used for the treatment of haemophiliacs.

Box 14   Warfarin

One of the features of a number of the proteolytic enzymes of the blood clotting cascade is that their activity depends on the presence of calcium ions, Ca^{2+}, tightly bound to the enzyme protein. This is not particularly unusual but what is special is the way in which these proteins trap the calcium ions. In Topic 4 and Fig. 4.1 we met the amino acid glutamic acid, one of the 20 that commonly make up proteins and noted that it has a negatively charged carboxyl group in its side-chain. The blood clotting factors contain a special amino acid, γ-carboxyglutamic acid, in which there is a second carboxyl group on the same carbon atom. This produces a sort of chemical 'claw' that is very effective at grabbing and holding Ca^{2+}. There has to be a way to make these 'claws', and it employs an enzyme that requires vitamin K as a cofactor.

This very specific involvement of vitamin K in clotting has opened up the possibility of pharmacological intervention in two rather different ways. The compound warfarin is a chemical analogue of vitamin K (Fig. 54.1) and was introduced as rat poison. With their clotting defence removed, the rodents die through internal bleeding. In more recent times, precisely the same compound has been widely used as a prescription drug, particularly for the elderly, where there is felt to be a serious risk of thrombosis. Clearly the use of such a compound is a delicate balancing act and control of dosage is critical.

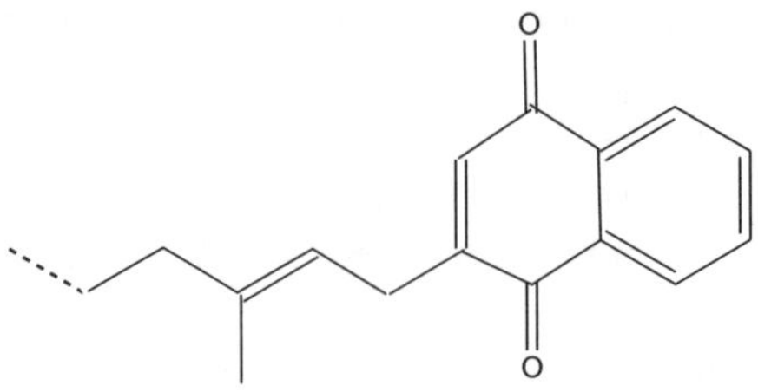

Vitamin K

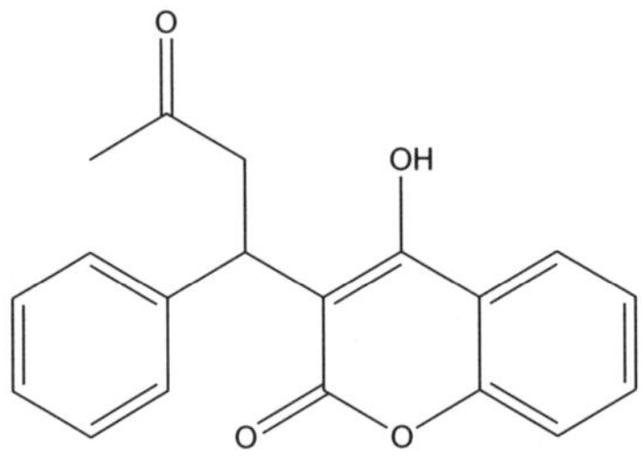

warfarin

Figure 54.1 Warfarin and vitamin K. The dashed line in the vitamin K structure indicates a long hydrocarbon chain (20 carbon atoms in the side chain altogether). There is sufficient similarity in the ring structures of the two compounds for warfarin to act as a molecular mimic, thus counteracting the action of vitamin K (see Appendix 14).

Self-test MCQs on Topic 54

1 Which one of the following plays a key role in initiating blood clotting?

(a) Red blood cells

(b) Platelets

(c) Lymphocytes

(d) Cholesterol

2 Thrombin plays which one of the following functions in blood clotting?

(a) It converts fibrinogen into fibrin.

(b) It forms the main protein component of the clot.

(c) It activates prothrombin.

(d) It prevents clotting getting out of control.

3 Aspirin is frequently given as a prophylactic drug. This is in order to do whioch if the following?

  (a) Relieve pain in the event of angina.

  (b) Lessen the risk of haemorrhage.

  (c) Decrease the tendency for clot formation.

  (d) Lower blood cholesterol.

4 Tissue plasminogen activator, TPA, is useful for which of the following?

  (a) Helping to dissolve blood clots.

  (b) Helping to curtail haemorrhagic events.

  (c) Initiating the clotting cascade.

  (d) Controlling blood pressure.

SECTION 6

Appendices

APPENDIX 1

pH and neutrality

As we have mentioned in Chemistry III, the concentration of the hydrogen ion in solution can vary over such wide ranges that it is not always convenient to talk about it in the usual units, mM, μM, etc. Instead, the convention is to assign a number, so that, if the concentration of H^+ is 10^{-9} M, for example, the pH is said to be 9. Note that, because the convention drops the minus sign in the exponent (e.g. the minus in 10^{-9}), the pH value gets *higher* as the H^+ concentration gets *lower*. Low pH means acid; high pH means alkaline.

The formal mathematical definition that describes this behaviour is that the pH value is minus $\log_{10} [H^+]$, with $[H^+]$ being expressed as molar concentration. In our example, $\log_{10} 10^{-9}$ is -9, and $-(-9)$ is $+9$. Hence, the pH is 9.

Why do we take pH 7 as the measure of 'neutral' pH, neither acid nor alkaline? The answer lies in the behaviour of water, H_2O. Water spends most of its time as the intact, undissociated molecule implied by the formula. A very small proportion, however, will dissociate to give hydrogen ions (protons), H^+, and hydroxyl ions, OH^-. The extent of this separation into ions is restricted by the need to obey an equilibrium constant (see Appendix 4), which dictates that:

$$[H^+] \times [OH^-] = 10^{-14} M^2$$

Now, if the solution is neither acidic nor alkaline, as should be the case in pure water, the concentrations of H^+ and OH^- should be exactly equal. If they have to multiply together to give $10^{-14} M^2$, then each one must be equal to the square root of $10^{-14} M^2$, which is 10^{-7} M. So the pH at neutrality is $-\log_{10} 10^{-7}$, which is 7. pH values above 7 are alkaline, therefore, and pH values below are acidic. We think of the physiological pH as being just above 7, but in particular locations it can be quite different. In the stomach, for example, the gastric juice is at a pH of below 2. Inside individual cells there are organelles called lysosomes (which you could almost think of as the stomach of the cell!) and these also have a very low internal pH of about 5.

Pain-Free Biochemistry   Paul C. Engel
© 2009 John Wiley & Sons, Ltd

APPENDIX 2

Crystallography

Many simple compounds can readily be persuaded to form crystals, solid forms with very regular and beautiful shapes. The reason they form these regular shapes is that the atoms or molecules that make them up are lined up in precise regular arrays. In the early years of the twentieth century, Sir William Bragg discovered that shining X-rays at crystals of sodium chloride (the little cubes that make up table salt) gave a regular pattern of 'diffraction' of the X-rays, leading to spots that could be captured on photographic film. It was possible to work backwards from the pattern to calculate and predict the regular structure that must have caused it. In the case of sodium chloride, Bragg caused a sensation by showing that the Na^+ and Cl^- ions (see Chemistry II) were arranged in an alternating lattice pattern so that each Na^+ was linked to several Cl^- ions and each Cl^- to several Na^+. The big-wig chemists of the time clearly had in mind a 'till-death-do-us-part' arrangement for each NaCl molecule, and so Bragg's ionic commune came as a shock!

Later on, in the 1920s, there was a big controversy over James Sumner's suggestion that enzyme proteins could be crystallised, the general view being that proteins were gooey substances ('colloids') that would never behave in such a regular and respectable fashion! However, Sumner was right, and in due course, J.D. Bernal and his pupils started shining X-rays at crystals of various biological molecules. Among them was Max Perutz, who attempted the seemingly impossible task of solving the structure of a whole protein, myoglobin, an oxygen storage protein from (whale) muscle. This led to the publication, in 1957, of the first ever solved protein structure, showing a complicated folded 3-D shape (Fig. A2.1). Despite the complex shape, the myoglobin molecules, just like the sodium and chloride ions, were lined up in absolutely regular arrays, so that they too gave a solvable pattern. Perutz's heroic task took many years. Today, with massive computer power to handle the huge number of calculations, and robotic devices to collect the mountains of data, it is possible to solve a structure in days.

Pain-Free Biochemistry   Paul C. Engel
© 2009 John Wiley & Sons, Ltd

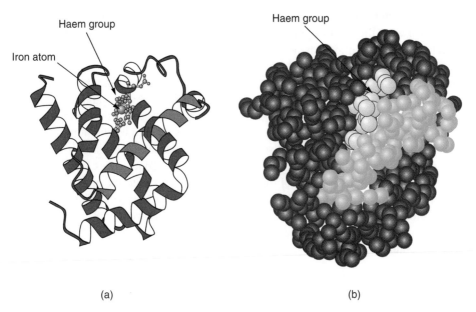

Haem group

Iron atom

Haem group

(a)   (b)

Figure A2.1 The structure of myoglobin, the first protein to have its 3-D structure solved by X-ray crystallography (1957). The first image, a so-called 'ribbon diagram' emphasises the course of the polypeptide chain that makes up the molecule and also clearly shows the regular 'secondary structure' – the long corkscrews that dominate this structure are α-helices. A true picture is given by the space-filling representation on the right, in which all the amino acid sidechains are put onto the backbone. One helix is picked out in blue to help you relate the two representations of the molecule. Also picked out, in grey, is the haem cofactor, which has an iron atom at its centre and enables the myoglobin molecule to capture and release oxygen. The same cofactor is found in haemoglobin in our red bood cells. Reproduced from Berg, Tymoczko and Stryer, *Biochemistry*, W.H. Freeman, New York, with permission of the publisher.

Apart from the sheer challenge and interest of solving the puzzle and finding out what a protein molecule actually looks like, does this have any relevance for medical science? Emphatically yes. First of all, if we think of each protein molecule (enzyme, antibody, hormone receptor, muscle protein, etc.) as a tiny machine, it is very difficult to understand how such a machine works if you do not know what it looks like. We now have structures for thousands of different proteins available in international databases. More specifically, in medical practice we very often want to interfere with the machinery and, if possible, do so in a way that is precise and effective. In other words, we want to be able to make drugs that hit their target and do not do too much damage in the way of undesirable side effects. There have been three main ways of approaching this. The first was to rely on the wisdom of our ancestors who discovered the power of salicylic acid from willow bark (leading to aspirin), digitalis from the foxglove, atropine from belladonna, morphine from the poppy and so on. They did not know how they worked; they just knew that they did work! The second approach, used by drug companies for much of the twentieth century, was to make thousands of chemical compounds and try them out to see what they did in pharmacological test systems – tedious, but ultimately effective, even though it was not based on special insight. The third approach has only become possible since we have had a proper insight into the body's chemistry and detailed structures provided by crystallography. If you can accurately identify your target – say the receptor for a neurotransmitter in the brain – and if you know its detailed structure, then you may be able to apply intelligent design to making the right compound in the first place. A good example is offered by the hunt for drugs to control AIDS. Once it was realised that AIDS is caused by a virus, it was soon apparent that a good target would be the enzyme that the virus uses for processing its own 'coat' protein. Solution of the structure of this protein (the enzyme) offered powerful insight to help in the design of drugs to block its activity. The speed with which biochemistry has produced a cocktail of drugs to control the menace of AIDS means that HIV-positive people can look forward to many years of health instead of a death sentence.

Protein forces, secondary structure and folding

As explained in Topic 35, the protein synthetic machinery just turns out a linear chain, like a bootlace or a string of spaghetti, and yet all the strings of spaghetti somehow find their way to the correct structure, complicated and 3-D. An American biochemist in the 1960s did a theoretical calculation, showing that, if a protein explored all the possibilities open to it, it would be likely to take a time longer than the age of the Earth (4000 million years) to find the right structure. Around the same time, another American biochemist, Christian Anfinsen, showed for one particular protein that it could fold up correctly over a period of hours. This is an experiment we can do because there are chemicals available, such as urea, that are able to unfold proteins completely, i.e. all the way back to the 'spaghetti' stage or the so-called 'random coil'. One can dilute out the urea and watch what happens, and many proteins can refold even faster than in Anfinsen's experiment, regaining their biological activity in minutes or even seconds. Since it does not, after all, take 4000 million years, presumably the protein molecule is smarter than to explore all the myriad hypothetical possible shapes. Instead, it must have built-in features that tend to steer it down the right path.

Broadly speaking we do know the main forces that influence and ultimately maintain protein structure.

1 Perhaps the easiest to understand, but not the most important, is **electrostatic** interaction, i.e. negatively charged side-chains will tend to be attracted to positively charged ones.

2 A powerful influence is **hydrophobic** interaction, discussed in Chemistry VIII. In the context of proteins it tends to mean that amino acid side-chains that are

purely hydrocarbon (e.g. phenylalanine and leucine in Fig. III.1) will tend to bury themselves together and away from the surrounding water. Protein chemists talk about the 'hydrophobic core' of a protein molecule. Equally, 'polar' side-chains with charges, —OH groups, etc. are more likely to be on the surface. As well as being a global effect, the hydrophobic forces may well exercise a local influence, leading particular bits of the amino acid sequence to fold up first and steer the overall process.

3 **Hydrogen bonds** are another major influence. What are they? If a group like an —OH or an —NH comes close to, say, the oxygen atom of a carbonyl group, $\rangle$C=O (see, e.g., Chemistry VI), the hydrogen atom seems to show a degree of infidelity! It is drawn towards the handsome oxygen atom with its unshared 'lone pair' of electrons, and while it does not go all the way and actually desert its bonded partner, it hangs around with the interloper. This attraction is not individually very strong, but in proteins the actual backbone structure that makes up

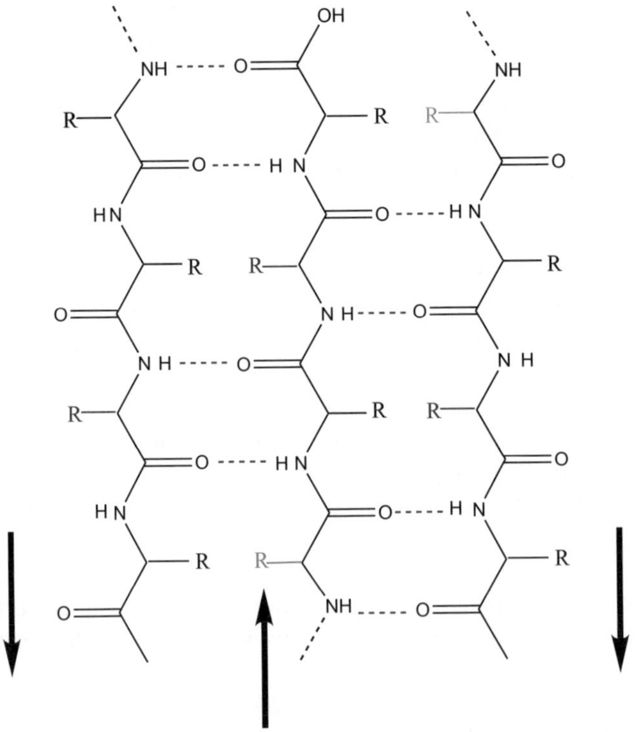

Figure A3.1 Strands of 'beta sheet'. The two strands of polypeptide lying alongside one another are able to form multiple hydrogen bonds, stabilising the structure. The adjacent strands here are 'anti-parallel' (running in opposite direction). It is also possible to have parallel strands, but the linking hydrogen bonds are then slightly bent (and therefore weaker) unlike the structure shown, in which the H-bonds are perpendicular to the strands.

the polypeptide chain has an underlying regularity: every amino acid in the chain contributes a carboxyl group, which becomes the $\rangle$C=O of the peptide bond, and also an amino group, which becomes an —NH. If now we stretch out a piece of polypeptide, it will present a regular pattern of $\rangle$C=O and —NH groups either side. If we bring two such strands side by side, the $\rangle$C=O groups on one can index with the —NH group of the other (Fig. A3.1) producing a structure with many hydrogen bonds, all reinforcing one another. In fact, several strands can come alongside to form a sheet. This is a very strong structure known as **beta sheet**, and pieces of beta sheet are found in many, if not most, proteins. Silk is an example of a protein made up entirely of beta sheet. This is not the only

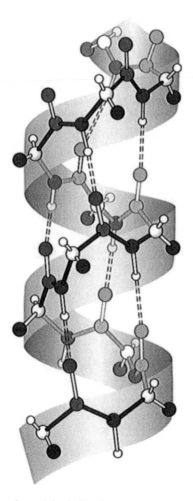

Figure A3.2   The structure of an alpha-helix. Oxygen atoms are shown in red and nitrogens in blue. In this spiral arrangement of the polypeptide chain, the C=O and N—H bonds are approximately parallel to the helix axis and able to form multiple H-bonds, shown here as double dashed lines.

regular structure stabilised by hydrogen bonds, however. The famous chemist Linus Pauling spotted, before anyone had solved a real structure of a protein, that you could also twist models of polypeptides into a spiral staircase pattern where again multiple hydrogen bonds could form the vertical supports of the staircase (Fig. A3.2). This structure, the **alpha-helix**, is also found in most proteins, and indeed the long sausages that make up the myoglobin structure (Fig. A2.1 in Appendix 2) are all stretches of alpha-helix. These regular, recurring structural patterns, which are found in many quite different proteins, are grouped together as features of secondary structure.

4 The previous three forces we have listed are all non-covalent. Some proteins, however, use more rigid struts in the form of covalent links. Most commonly these are **disulphide links** between two cysteine side-chains. Cysteine is one of only two amino acids containing sulphur as well as C, H, O and N in its molecule, and in the case of cysteine, the S appears in an —SH (sulphydryl or thiol) group at the end of the side-chain. —SH groups are chemically reactive and very easily oxidised, and if two of them come together they can be jointly oxidised to give an —S—S— link, the disulphide link. This structure is missing from many intracellular proteins, but proteins designed for a rugged existence, e.g. the enzymes of our digestive tract, frequently are reinforced by a number of disulphide bonds.

In this context it is worth mentioning a different kind of covalent reinforcement that is found in a class of proteins called **collagens**. These are the main proteins in bone, cartilage and connective tissue. Being structural proteins in our bodies, collagens need to be strong, and their molecules are reinforced by special enzymatic mechanisms that create crosslinks between lysine side-chains.

APPENDIX 4

Equilibrium constant

If we imagine a chemical reaction, in which two substances A and B react together to give two products C and D, as discussed in Chemistry IV, we should normally expect the reaction to proceed at a rate that is proportional to the concentration of A and also proportional to the concentration of B. After all, if, say, we double the concentration of A, then we are doubling the chances of a molecule of A bumping into a molecule of B and so also the chance of A and B reacting.

We can state this mathematically by writing:

$$\text{rate of forward reaction} = k_1[A][B]$$

(k_1 is a constant, called a rate constant, and the square brackets denote the concentration of the substance inside the brackets.)

If there are also some product molecules, C and D present (and there are bound to be once the forward reaction gets going, whether or not there were any to start with), we can also think about the rate of the reverse reaction to make A and B from C and D, and state that:

$$\text{rate of reverse reaction} = k_2[C][D]$$

If we start off with only A and B in solution, then the reverse reaction rate will be zero initially, but it will increase and the forward rate will dwindle as [A] and [B] decrease and [C] and [D] increase. Eventually a point will be reached where the two rates exactly balance, so that:

$$k_1[A][B] = k_2[C][D]$$

Pain-Free Biochemistry   Paul C. Engel
© 2009 John Wiley & Sons, Ltd

This equation defines equilibrium and it is true whatever the starting concentrations may have been. It can be rearranged to read:

$$\frac{[C][D]}{[A][B]} = \frac{k_1}{k_2} = K_{eq}$$

K_{eq} is the equilibrium constant. Let us suppose that for our particular reaction $K_{eq} = 16$, and that we start off with 10 mM A, 10 mM B, and no C or D. We can easily prove that at equilibrium the starting reactants will be 80% converted into products so that we end up with 2 mM A, 2 mM B, 8 mM C and 8 mM D:

$$((8 \times 8)/(2 \times 2) = 64/4 = 16).$$

So this equilibrium lies over to the right, but not all that far over. It might be that we are keen to convert all the B to D via this reaction. If so, we could opt to push the equilibrium over by leaving starting [B] the same at 10 mM but increasing the starting concentration of A 10-fold to 100 mM. Using the same equilibrium constant (we have to!), we can work out that the final concentrations of C and D will now be approximately 9.93 mM; [B] will be down from 10 to 0.07 mM, but in the case of A, although its concentration also must be down by 9.93 mM (molecules do not disappear into thin air or appear from nowhere!), this still leaves 90.07 mM. So by increasing [A] 10-fold, we have gone from only 80% conversion of B into D up to over 99% conversion. Thus, we can push a reaction by increasing the concentration of one or more of the reacting substances.

Could we turn this reaction round to make B from D and again achieve almost 100% conversion? This time, if we start with 10 mM C and 10 mM D, we shall end up with exactly the same equilibrium mixture as the first one above, i.e. we shall get only 20% conversion of D into B and C into A. If again we go up to 100 mM with the other substrate, C, this time it results in only 53% conversion of D into B (5.3 mM B and 4.7 mM D left, roughly, at equilibrium). This is because the scales are tipped to start with, since K_{eq} is 16 rather than 1 or 0.1. Nevertheless, the reaction can be pushed still further if we try hard enough. In this case going up to 1000 mM C, still with 10 mM D, results in 88% conversion of D into B. (To solve these for yourself you need to set the amount of each product as x mM and solve the quadratic equation, but if you just want to test whether the numbers work, you can easily just plug each set into the expression for K_{eq} and see if it comes to approximately 16.)

This example emphasises the effect of an equilibrium constant in determining how far a reaction can go: it is important to remember that reactions can only ever proceed towards equilibrium and cannot go past that point. This is fundamental to understanding the whole point of 'energy metabolism' in our bodies. Our cells and tissues need to be able to switch between breakdown and synthesis of a substance (say glucose) without having to rely on or wait for massive swings in concentration such as we have just explored in our calculations. This is why, as we shall discover in Topics 25–27, synthetic pathways do not simply reverse the corresponding breakdown pathways. The cell must somehow arrange things so that all the equilibria are favourable in whichever direction it wants to proceed. This is where ATP comes in again and again.

APPENDIX 5

Phosphorus, phosphoric acid and phosphate esters

The element phosphorus, with the symbol P, has a valency of 5. In biochemistry we meet this element again and again in combination with oxygen, as the phosphate ion.

Phosphoric acid, H_3PO_4, or, to show its structure more clearly, $O{=}P(OH)_3$, is a 'tribasic' acid. This means that it is capable of losing 1, 2 or 3 protons (from the three —OH groups) to form ions with 1, 2 or 3 negative charges ($H_2PO_4^-$, HPO_4^{2-} and PO_4^{3-}) (Fig. A5.1).

Being an acid, phosphoric acid forms esters with alcohol compounds (see Chemistry VI and VII), and those regularly encountered in biochemistry are sugar phosphates such as G6P. Also containing a sugar phosphate structure are the nucleotides such as AMP, adenosine monophosphate. In these molecules, the sugar is linked to phosphate through one position (position 5 of the five-carbon sugar ribose, in the case of AMP; another position, 1, is linked to adenine). You may therefore sometimes see its name written more fully as adenosine 5′-monophosphate. The 5 tells us which sugar position carries the phosphate, and the little ′ sign tells us that it is indeed the sugar carrying the phosphate and not the adenine structure, which is separately numbered starting from 1.

Since phosphate is tribasic, rather like the sugar structure, it can combine at more than one position, and so we find that AMP can link a second phosphate onto the first to make ADP and a third one onto the second to make ATP. Equally important, in the nucleic acids, RNA and DNA (Topic 33), we find that the phosphate linked to the 5′ position of one sugar can link to the 3′ position of another, which in turn can attach another phosphate at its 5′ position and so on, making a chain. This linkage (Fig. A5.2), in which a single phosphate links two sugar units, is called a phosphodiester linkage.

Pain-Free Biochemistry   Paul C. Engel
© 2009 John Wiley & Sons, Ltd

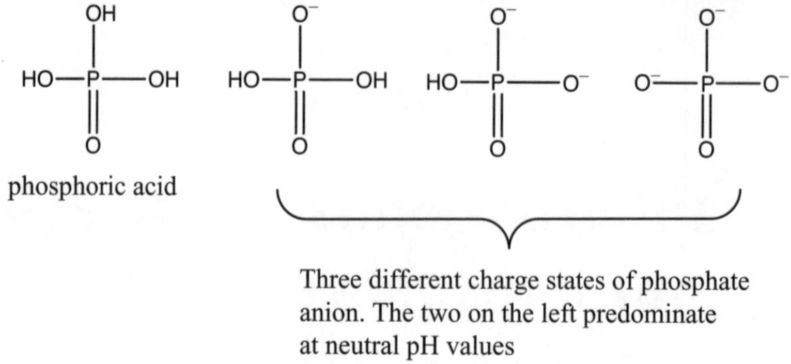

phosphoric acid

Three different charge states of phosphate anion. The two on the left predominate at neutral pH values

Figure A5.1 Different charge states of phosphate ion/phosphoric acid. Depending on the surrounding pH, phosphoric acid is able sequentially to shed up to three protons.

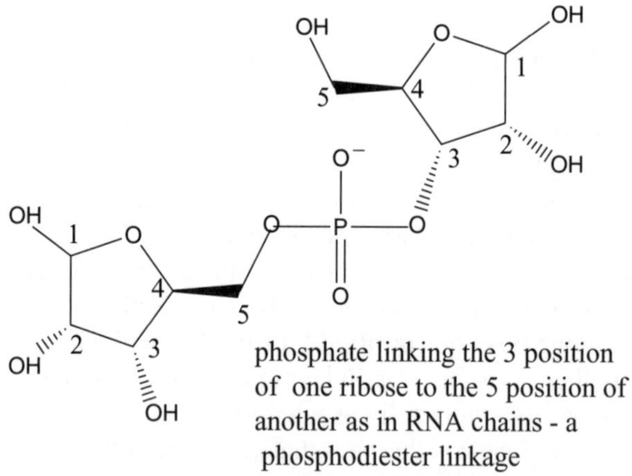

phosphate linking the 3 position of one ribose to the 5 position of another as in RNA chains - a phosphodiester linkage

Figure A5.2 Phosphodiester linkage.

APPENDIX 6

Coenzymes, cofactors and prosthetic groups

Although the 20 amino acids and complex folding patterns open up a range of chemical possibilities to proteins, there are also many limitations. In order to widen the range of chemistry available to enzyme catalysts, we make use of auxiliary compounds particularly well suited to certain types of reaction (Fig. A6.1). Thus NAD^+, $NADP^+$ and FAD are nucleotide cofactors based on the structures of nicotinamide and riboflavin, and are used for oxidation–reduction reactions. Biotin is used for carboxylation, pyridoxal phosphate is used for transamination and decarboxylation and so on. As mentioned in Box 3 (Topic 10), most of these are derived from vitamins in the diet.

An important point to appreciate in relation to these biochemical compounds is that, being present in only limited quantities in the cell, instead of piling up as net products of a metabolic pathway, they have to turn over and over, being recycled many times. A large flow-through of bulk substrates like glucose or fatty acids is handled by relatively tiny amounts of cofactors.

Another potential puzzle lies in the difference between the modes of operation of a coenzyme like NAD^+ and a prosthetic group like FAD, both of them oxidation–reduction cofactors. The fundamental difference is that a prosthetic group forms a permanent part of the equipment of the enzyme, whereas a coenzyme like NAD^+ arrives and departs just like any other substrate. In an oxidation carried out with the help of FAD, the reduced FAD would just have to wait for another substrate molecule to arrive to put it out of its misery and reoxidise it. In a similar reaction with NAD^+, the NADH, once produced, would simply leave and find its own way to be reoxidised with the help of another protein partner.

Pain-Free Biochemistry   Paul C. Engel
© 2009 John Wiley & Sons, Ltd

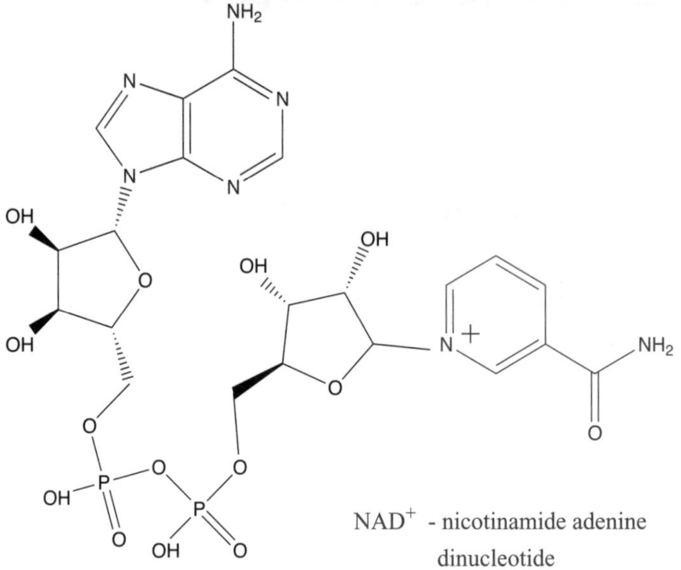

NAD$^+$ - nicotinamide adenine
dinucleotide

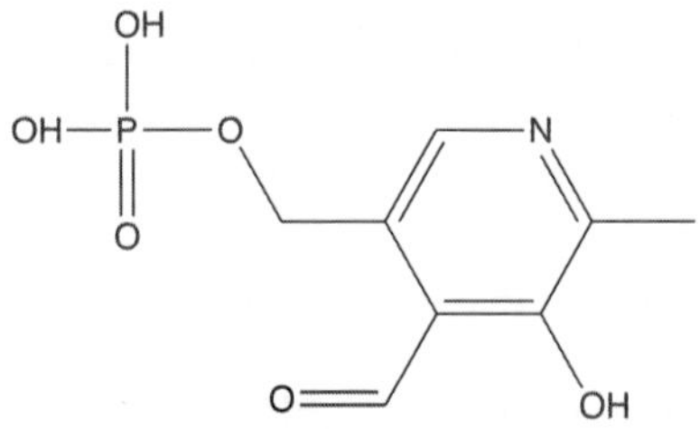

pyridoxal phosphate

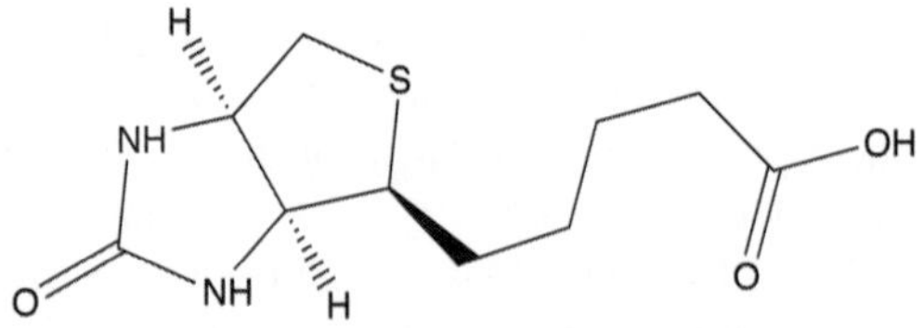

biotin

Figure A6.1 Structures of three of the vitamin-derived cofactors. NAD$^+$, like several other cofactors, has a 'handle' consisting of adenine (blue) and ribose phosphate (black), but the actual chemistry of oxidation and reduction is done by the nicotinamide (red), which is derived from the vitamin nicotinic acid or niacin. Note the positive charge on the ring nitrogen in the oxidised form, NAD$^+$, which is shown here. Pyridoxal phosphate, derived from vitamin B$_6$ carries out its chemical contribution via the aldehyde group (mauve). Biotin likewise contributes its own characteristic chemistry, with the nitrogen atom, shown in blue, readily able to pick up CO$_2$ as a carboxyl group.

APPENDIX 7

Coenzyme A

CoA is similar in some ways to other coenzymes we have encountered but rather different in others. It is derived, like a number of others, from a B vitamin, pantothenic acid. It also has a nucleotide structure, incorporating adenine and the sugar alcohol, ribitol. Functionally, however, it behaves rather differently. When it is doing its job, e.g. in the β-oxidation of fatty acids (Topic 20) or assisting the entry of acetyl groups to the Krebs cycle (Topic 14), it is neither part of the enzyme like, say, FAD nor a separate substrate like NAD$^+$. Instead, it is *part* of a substrate. In acetyl CoA or in a long-chain fatty acyl CoA, the acyl chain is attached to the sulphur of CoA and this provides chemical activation, making something that would otherwise be chemically sluggish into something chemically reactive. Synthetic chemists do something very similar and equivalent when they convert an unreactive carboxylic acid to a highly reactive acyl chloride. At the time the Krebs cycle was discovered in the 1930s, it still was not clear exactly how the C2 units entered to make citric acid. It clearly was not acetate itself that was the substance involved, but what was it? To cover for this gap of ignorance, biochemists had to use the term 'active acetate' for a while. Eventually, in 1945, Fritz Lipmann discovered CoA, so-called because it was required in order to bring about the *a*cetylation of drug substances. It quickly became apparent that its significance was much wider.

If it arises from pyruvate (Topic 14), the formation of acetyl CoA is pushed by the energetically favourable oxidative decarboxylation reaction. If, on the other hand, one starts from acetate, or similarly from a long-chain fatty acid waiting to enter the fatty oxidation spiral, the formation of the acyl CoA involves an 'activation' reaction in which ATP is split.

$$RCOOH + CoASH + ATP \rightarrow RCOSCoA + AMP + \text{pyrophosphate}$$

Pain-Free Biochemistry   Paul C. Engel
© 2009 John Wiley & Sons, Ltd

One thing that confuses students is the way that we sometimes write CoA and sometimes CoASH. They are one and the same thing, and both are shorthand representations rather than real structures. It is just that in representing an actual reaction we tend to emphasise the chemical group in CoA, the thiol group (−SH), that is actually directly involved in the reaction and to which the acyl group becomes attached. There is *no* special significance in the fact that in CoASH we write the SH to the right and in RCOSCoA the S is to the left!

APPENDIX 8

Krebs cycle and evidence for a catalytic reaction sequence

In the 1920s and early 1930s, people were trying to puzzle out how pyruvate produced in the breakdown of glucose got further oxidised, ultimately to CO_2. In search of clues, they looked to see exactly what simple organic compounds cells could oxidise ('respire'). It turned out that, while many compounds could be oxidised, there was a select group that seemed to be able to be oxidised remarkably rapidly. These included such things as malic acid, succinic acid, oxaloacetic acid, etc. When careful measurements were made, it seemed that with these, the observed oxidation was not just the oxidation of what had been added. If you added an exact amount of, say, succinic acid, you could calculate exactly how much CO_2 it should produce. In fact, much larger amounts of CO_2 resulted! So what was happening? It emerged that these compounds were somehow able to speed up the oxidation of glucose, pyruvate, etc. If this seemed unlikely, a powerful piece of evidence came from the fact that a compound called malonate, which directly blocked the oxidation of succinate (Fig. A8.1 and see also Appendix 14), *also* blocked the oxidation of pyruvate or glucose! But how could these compounds have anything to do with the oxidation of pyruvate? Hans Krebs solved the puzzle by showing that there was an enzymatic step that converted four-carbon oxaloacetic acid to six-carbon citric acid by adding a two-carbon unit from 'active acetate' (see Appendix 7), which was some years later identified as acetyl CoA, produced, as we have seen, by the oxidation and decarboxylation of pyruvate.

This discovery was in the end even more important than it at first seemed, since it became clear in due course (1) that fatty acids also gave rise to acetyl CoA in their oxidation (Topic 20), (2) that several amino acids also did so (Topic 23) and (3) that some amino acids were converted directly into one or other of the Krebs cycle

Pain-Free Biochemistry   Paul C. Engel
© 2009 John Wiley & Sons, Ltd

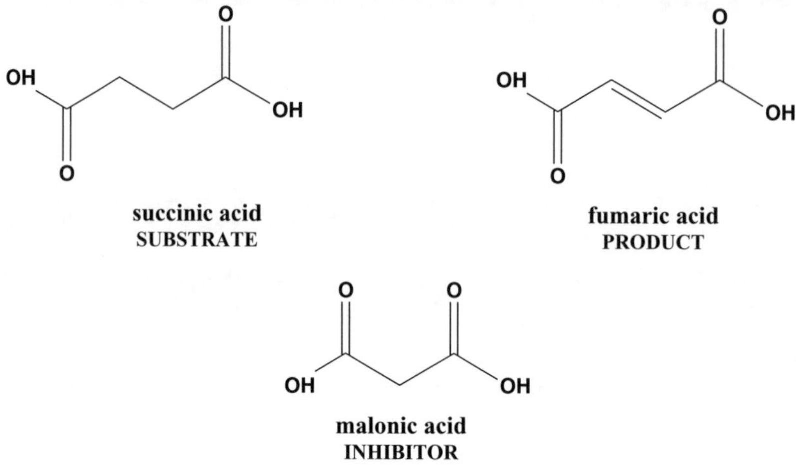

succinic acid
SUBSTRATE

fumaric acid
PRODUCT

malonic acid
INHIBITOR

Figure A8.1 Succinate and malonate, a competitive inhibitor of succinate oxidation in the Krebs cycle.

intermediates. Thus, the Krebs cycle came to be known as a 'common terminal pathway of oxidation'.

The cycle works, on a grand scale, in rather the same way as a coenzyme does in a single reaction: the compounds are constantly regenerated and speed up the oxidation of two-carbon fragments without being used up in net terms themselves. The whole set of substrates and enzymes, taken together, catalyses the conversion of acetyl groups into CO_2 and water, and of course does it in such a way that reduced cofactors are produced in order to drive formation of ATP (see Topic 14).

APPENDIX 9

Knoop's experiment pointing to β-oxidation of fatty acids

Knoop's famous experiment was published in 1904 and was done with dogs. It was important not just in the study of fatty acid oxidation, but much more generally, because it introduced a novel method to biochemistry, the method of chemical labelling. Knoop fed his dogs fatty acids that had a phenyl ring at the end furthest from the —COOH group, and he collected their urine to see if he could find what the dogs had managed to do with these compounds. The metabolic products he found were attached to the amino acid glycine (this is one of the body's common ways of dealing with potentially toxic compounds, making them more soluble and more easily excreted by the kidney). Knoop had used some fatty acids with an even number of carbon atoms in their chain and some with odd numbers. He found that he had two types of end product, but which one he found depended on which fatty acid he used. Odd chains gave him benzoyl glycine. Benzoic acid has just one carbon attached to the phenyl ring in a —COOH group. When he fed the dogs fatty acids with even numbers of carbons in the chain they produced no benzoyl glycine. Instead, they made phenyl acetyl glycine, with two carbons attached to the phenyl ring (—CH$_2$COOH).

This experiment proved

1 that fatty acids could be broken down from the —COOH end; all that was left was a last bit attached to the phenyl ring at the other end;

2 that they could not be broken down one carbon unit at a time. If they were, then the first cycle would convert odd into even or vice versa and thereafter everything would progress in the same way so that odd or even would end up giving the same products.

Pain-Free Biochemistry   Paul C. Engel
© 2009 John Wiley & Sons, Ltd

Knoop proposed that fatty acid chains were broken down two carbons at a time, and even went on to propose a theory of β-oxidation, suggesting that if the β-carbon could be converted from CH_2 into $)C=O$ then you could split off acetate and leave a fatty acid two carbons shorter. As we have seen in Topic 20, he was very close to the truth, and at this point was well ahead of the state of play on glucose oxidation. However, over the next few years the involvement of sugar phosphate in glucose metabolism was discovered and by the 1930s the pathway and enzymes of glycolysis were more or less worked out, whereas the same was not true for fatty acid oxidation until the early 1950s. This remarkable gap was due to our ignorance of CoA. Even when it was discovered, it was not at first readily available for research, making it quite difficult to hunt for the enzymes that handle CoA-containing substrates.

Modern biochemists would not use Knoop's method and would say that he had taken a risk that the dogs' metabolism might object to the phenyl labels and either not handle them at all or perhaps handle them in a different way from the metabolism of normal fatty acids. Nowadays, we would be able to fool our metabolism much more subtly and completely by using isotope labels (see Chemistry X).

APPENDIX 10

Isoenzymes

The availability of increasingly sensitive analytical techniques for separating proteins, notably electrophoresis (see also Appendix 15), in the 1960s gradually revealed that more and more of the familiar mainstream enzymes, e.g. those of the glycolytic pathway occur in more than one form in the body. It is now clear that very often there are separate genes encoding these isoenzymes. Often there can be more than one form of an enzyme in the same tissue, but, even so, one of the obvious aspects of isoenzymes is that different tissues tend to have different distributions of the different forms.

A striking example is that of the enzyme LDH. This enzyme, which we met in Topic 12, interconverts pyruvic acid and lactic acid. There are, however, three distinct LDH genes. One encodes a form of LDH, which seems to be found only in the testis. The other two LDH proteins are much more widespread, but, basically, one seems to be tailored for an anaerobic role, making lactic acid in tissues like skeletal muscle, whereas the other is better adapted to the reverse, aerobic conversion of lactic acid back to pyruvic acid (see Appendix 15 for the way in which this shows up analytically).

Another interesting case is the conversion of glucose and ATP into G6P and ADP (at the top end of glycolysis). One form of the enzyme, found in many tissues and commonly called hexokinase, gets into its stride with very low concentrations of glucose and is halfway to going flat out when the glucose reaches about 10 μM. The other form, catalysing the same reaction and called glucokinase, is mainly found in the liver and only gets moving at concentrations of glucose a thousand times higher. Ones immediate reaction is to think that perhaps one enzyme is good at its job and the other is bad, but in fact both are good! How can this be? The answer is that both are adapted for their particular role. Hexokinase 'expects' to encounter glucose concentrations rigidly controlled by the blood glucose level and membrane glucose transporters. Glucokinase, on the other hand, is there to deal with the tide of

Pain-Free Biochemistry Paul C. Engel
© 2009 John Wiley & Sons, Ltd

food-derived glucose that will come in through the hepatic portal vein draining the intestine and giving, periodically, a far higher intracellular concentration in the liver.

Finally, one should mention several other enzymes also associated with glycolysis, where there may even be three forms, but in particular there are often different forms in nervous tissue and in muscle, even though both tissues are presumably using the enzyme for the same glycolytic purpose. Reasons are not fully understood, but one might speculate that either these enzymes have to interact with other cell components (e.g. muscle fibres?) or else they have to respond to tissue-specific chemicals.

APPENDIX 11

Genetic code

UUU Phe	UCU Ser	UAU Tyr	UGU Cys
UUC Phe	UCC Ser	UAC Tyr	UGC Cys
UUA Leu	UCA Ser	UAA STOP	UGA STOP
UUG Leu	UCG Ser	UAG STOP	UGG Trp
CUU Leu	CCU Pro	CAU His	CGU Arg
CUC Leu	CCC Pro	CAC His	CGC Arg
CUA Leu	CCA Pro	CAA Gln	CGA Arg
CUG Leu	CCG Pro	CAG Gln	CGG Arg
AUU Ile	ACU Thr	AAU Asn	AGU Ser
AUC Ile	ACC Thr	AAC Asn	AGC Ser
AUA Ile	ACA Thr	AAA Lys	AGA Arg
AUG Met*	ACG Thr	AAG Lys	AGG Arg
GUU Val	GCU Ala	GAU Asp	GGU Gly
GUC Val	GCC Ala	GAC Asp	GGC Gly
GUA Val	GCA Ala	GAA Glu	GGA Gly
GUG Val	GCG Ala	GAG Glu	GGG Gly

*As explained in Topic 34, the codon AUG for Met also has to double up as the START codon.

It is not at all important for you to learn the genetic code, but it is interesting to see how it works. The code is written in terms of the mRNA codons that will encode each amino acid. Thus, the corresponding DNA sequence in the gene is the 'complementary' base-pairing sequence, e.g. AAG in the mRNA encodes lysine, and it will itself have been encoded by TTC in the DNA since T base-pairs with A and C base-pairs with G.

The 20 amino acids are indicated by their standard three-letter abbreviations: Phe, phenylalanine; Leu, leucine; Ser, serine; Tyr, tyrosine; Cys, cysteine; Trp, tryptophan; Pro, proline; His, histidine; Gln, glutamine; Arg, arginine; Ile, isoleucine; Met, methionine; Thr, threonine; Asn, asparagine; Lys, lysine; Val, valine; Ala, alanine; Asp, aspartic acid; Glu, glutamic acid; Gly, glycine.

Note that there are 8 of the 16 boxes with just 1 amino acid in them, e.g. all 4 codons beginning with GU code for valine. Only Met and Trp have single, unique codons. Leu, Ser and Arg have six codons apiece. All the others have 2, 3 or 4.

APPENDIX 12

Different kinds of mutation

In Topic 36, we explored the fact that, despite its high accuracy, the DNA machinery can go wrong, and if it is the DNA replication that has gone wrong, this produces an error, a mutation, which is more or less permanent because once it is there it will be handed on from generation to generation – unless, of course, it is so damaging that the line of cells dies out or, by chance, another mutation corrects it ('reversion'). In practice, what will a mutation do to the protein that the gene encodes?

We can get a certain amount of insight by looking at a simple sentence made up of three-letter words:

THE FAT DOG BIT THE CAT

This is not a perfect analogy because it skips the coding that translates one sequence (DNA) into another (protein), but it still helps us to grasp some of the possibilities.

Suppose we have a 'mutation', a typo, that replaces the first A with an I. Then we get:

THE FIT DOG BIT THE CAT

This still makes perfectly good sense, perhaps better sense than the first version, but it is certainly giving us a different message. The DNA equivalent is a 'point mutation' swapping one base for another and the likely consequence is a swap of one amino acid for another. (Because of coding 'redundancy' there is also the possibility of a 'silent' point mutation and we shall come back to that shortly.)

Next, let us consider another type of mutation. Suppose that the machinery skips three letters, dropping I, T and D so that the message now reads:

THE FOG BIT THE CAT

Pain-Free Biochemistry   Paul C. Engel
© 2009 John Wiley & Sons, Ltd

This is no longer a very convincing account, but the words are still meaningful. In the same way, a mutation that takes out three DNA bases (or six, nine, etc.) will still code for protein, although it will be shorter and very likely faulty. This is a deletion. You can probably see that in the same way we could have an insertion:

THE FAT DOG BIT THE C<u>OW P</u>AT

Again the words still make sense but the sense is different and the sentence is longer.

On the other hand, going back again to the first sentence, suppose we have a single-letter deletion, say, of the first A. We know the rules; we have to look for three-letter words and so we write out:

THE FTD OGB ITT HEC AT. . .

Making a deletion (or an insertion) that is not a multiple of three letters immediately throws everything 'out of register', and this usually abolishes sense. At the DNA level it creates what is called a nonsense mutation, and it would lead to a protein with all the wrong amino acids beyond the mutation. It might still be approximately the right length but very likely not. This is because not only the amino acids are encoded by three base codons but also the punctuation to tell the machinery that it has reached the end of a gene is also similarly encoded. So it is very likely that, once the whole sequence has been thrown out of register, sooner or later one of the new, wrong codons will be a STOP codon, resulting in a truncated protein. Thus, for example, if the original mRNA sequence was

CUU UGC GUA ACG. . ..

this would encode (see Appendix 11) the amino acid sequence
    Leucine–Cysteine–Valine–Threonine. . . . .
    If a deletion removes the first C, then we have

UUU GCG UAA CG. . . . .

UUU now codes for Phe, GCG codes for alanine, but when we get to the next codon, UAA is a STOP codon, and so the new, mutant protein would end there.

One can find examples of all these kinds of mutations, but the most common are point mutations, and, if we look at a single gene across the human population, there are usually a number of sites where one finds two or more variants in the DNA. This may be either because the change of amino acid does not result in a disastrous change for the protein, or perhaps that the mutation is 'silent'. For example, if we have a codon AAA in the DNA, this will give complementary UUU in the mRNA, coding for Phe. If a mutation changes AAA to AAG, then this will result in UUC in the mRNA, but UUC still codes for Phe, which has two possible codons, and so, despite the mutation, there is no change in the protein.

APPENDIX 13

Restriction enzymes

The knowledge of individual genes and their sequences and the ability to carry out genetic profiling for clinical or forensic purposes depend very heavily on a remarkable group of enzyme tools. **Restriction enzymes** were originally discovered in bacteria, and they are so named because they restrict the range of bacteria that can be invaded by bacterial viruses. We now know that they are found in other biological species too. Their function is defensive. They chop up the DNA of invading life forms. As we noted in Topic 23 in relation to protein degrading enzymes, such potent weapons are potentially disastrous also for the cells that make them. These DNA degrading enzymes need somehow to be intelligent enzymes that can distinguish between the DNA of the invader and the invaded. The full story is beyond our scope here, but a crucial feature that has made restriction enzymes extraordinarily valuable as a laboratory tool outside the cell is their very remarkable specificity. We have seen elsewhere that protein degrading enzymes are quite fussy and often require one particular amino acid at the site where they cut the polypeptide chain; restriction enzymes very often require a precise sequence of nucleotide bases as many as six or even eight long. Thus, for example, the restriction enzyme EcoRI (the cryptic name reflects the fact that it is produced by the bacterium *Escherichia coli*) will only cut double-stranded DNA if it finds the exact sequence GAATTC. Even though there are only four bases to choose from at each position, one can calculate that, roughly, the chance that any particular stretch of DNA will have exactly this sequence is 1 chance in $4 \times 4 \times 4 \times 4 \times 4 \times 4$, i.e. 1 in 4096. This does not mean that every DNA fragment produced will be over 4000 bases long, but what it does mean is that most fragments will be fairly large and that therefore much fewer fragments will be produced than if the enzyme produced smaller fragments by cutting more frequently.

This means that with the right choice of restriction enzyme(s) it may be possible to cut out a gene intact, without the enzyme finding any cut sites in the middle. It

Pain-Free Biochemistry   Paul C. Engel
© 2009 John Wiley & Sons, Ltd

Example DNA Fingerprint

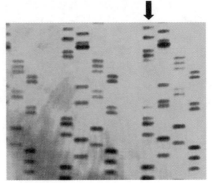

•Column in Gel corresponds to a person's DNA.

•It is called a person's DNA fingerprint.

•Row corresponds to DNA fragment size.

Figure A13.1   DNA identification. Variations in our DNA lead to different sizes of fragment when it is cut up with restriction enzymes. This is the basis of the DNA identification now so useful for forensic identification.

also means that samples of DNA can be cut into a limited number of fragments that can be separated by size, giving rise to a distinctive 'bar code' pattern (Fig. A13.1).

APPENDIX 14

Enzyme inhibition

We have seen throughout this book that the metabolic activities of different tissues at different times depend on the presence of individual enzymes, making it possible for the corresponding chemical reaction to occur at a physiologically realistic rate. In Topics 6 and 7, we saw how enzymes work. Our own cells regulate metabolic processes both by controlling the availability of enzyme proteins and by switching them on or off (Topics 39 and 42). However, it is also possible to decrease enzyme activity by administering foreign compounds. Depending on whether the effect is deliberate or accidental, and whether the intention is friendly or hostile, this can be the concern of toxicology or pharmacology. In either case, the process is termed 'enzyme inhibition' and the responsible chemical agent is an **inhibitor**.

In terms of drug action, we now know that many traditional drugs work by their action on enzymes, but also, now that biochemistry offers a detailed account of the chemical workings of different cells and processes, it is possible to contemplate rational drug design, which basically means selecting the right target in the cell and making a molecule to hit the target. Since the target may well be an enzyme, it is important to think about how this attack might work in practice.

First of all, we need to distinguish reversible and irreversible effects. An **irreversible inhibitor** typically will combine with the enzyme protein in a covalent way (Chemistry II), probably at the active site. What this does in effect is take enzyme molecules permanently out of action, so that the only way the cell can overcome the effect is to make some more enzyme protein, which, of course, takes time. There are a number of drugs that work like this, including aspirin (see Topic 49) and also a number of anti-depressant drugs. In view of their irreversibility, they are on the one hand effective but on the other hand potentially dangerous – controlling

Pain-Free Biochemistry   Paul C. Engel
© 2009 John Wiley & Sons, Ltd

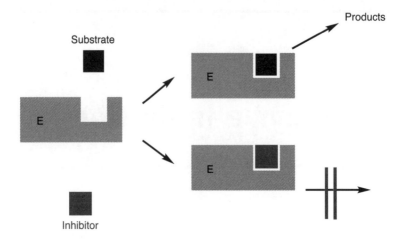

Figure A14.1 Competitive inhibition.

dosage becomes critically important in a situation where there is no short-term solution to an overdose.

A **reversible inhibitor**, on the other hand, does not react chemically with the enzyme molecule. Nevertheless, it will have a strong selective affinity for the surface
of the enzyme molecule and will tend to sit there, held by non-covalent forces. How
does such a compound inhibit the enzyme's activity? There are two broad possibilities. The most obvious one is that the inhibitor is a compound with a very similar
chemical structure to that of the enzyme's substrate (or one of its substrates if it
has more than one). In that case it will be like a cuckoo in the nest, sitting in the
active site instead of the substrate (Fig. A14.1). Since the two molecules are then
competing for access to the same site on the enzyme surface, this is called **competitive inhibition**. Obviously, just as the inhibitor can displace the substrate, so

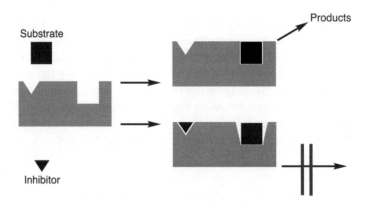

Figure A14.2 Non-competitive inhibition.

also the substrate can displace the inhibitor. One consequence of this is that if the inhibitor blocks a metabolic pathway, the resulting pile-up behind the block, like water behind a dam, may finally overcome the inhibition. An inhibition of this kind may, therefore, only be temporary. If, on the other hand, the inhibitor competes with a coenzyme (e.g. NAD^+) of which there is only a limited pool, there is no real possibility of a pile-up. This inhibition might be longer lasting, therefore, but a disadvantage is that the same inhibitor is likely to inhibit a dozen other enzymes that all use the same coenzyme!

An alternative possibility is that the inhibitor attaches to the enzyme somewhere else (possibly close to the active site but not at it), and, because proteins are flexible, it causes a distortion of the active site. As a result, the substrate may be able to attach to the enzyme as usual but without any catalytic outcome, i.e. the inhibitor has sabotaged the activity from a distance (Fig. A14.2). This second pattern is known as **non-competitive inhibition**. It may be harder to plan for in terms of rational design of drug molecules, but on the other hand it is likely to lead to a more persistent drug effect, since accumulation of the substrate will not displace the inhibitor.

APPENDIX 15

Electrophoresis to separate proteins

What is electrophoresis?

Biological fluids, such as blood plasma, are likely to contain large numbers of different kinds of proteins, each there to do a different job or occasionally there inappropriately, perhaps because of tissue damage. The clinical biochemist needs to examine such samples, if possible quickly and cheaply, and spot anything out of the ordinary that may indicate a medical condition. As explained in Topic 53, one can look for individual active proteins, e.g. enzymes, by using a specific measurement of their biological activity. However, it can be very useful in the first instance to have a look at the overall pattern. Electrophoresis is a very convenient way of doing this. It is an analytical technique that separates molecules on the basis of their different speeds of migration in an electric field. In the case of protein molecules, it makes use, first of all, of the fact that proteins have many ionisable side-chains (see Chemistry III and Topic 4). This means that unless, by chance, the negatives exactly balance the positives, each protein molecule is likely to carry a net electrical charge. In an electrical field, therefore, they will move: negatively charged proteins will move towards a positive electrode and vice versa. Since the charge will vary from protein to protein depending on its amino acid composition, different proteins will move faster or slower. In addition to the actual charge, another factor will also affect how fast each protein moves, namely its size and shape.

Cellulose acetate strips

Two distinctly different ways of carrying out electrophoresis have been popular. In the first, the proteins are applied in a thin line onto the surface of a dampened strip

Pain-Free Biochemistry   Paul C. Engel
© 2009 John Wiley & Sons, Ltd

of cellulose acetate, a thin, flexible, plastic-like substance. The two ends of the strip dip into two shallow tanks, which are separately wired up, one to the positive and the other to the negative terminal of a power supply. The electrical circuit is completed by the liquid in/on the cellulose acetate strip, and the current is carried by the movement of ions in solution. This includes the charged protein molecules, and so, although they all start off together in the applied biological sample, they gradually separate. Usually most of the proteins will have a net negative charge, but any that are positive will, of course, move in the opposite direction from the rest. At the end of the electrophoresis run there may be nothing to see, unless the sample includes coloured proteins like haemoglobin. Most proteins, however, are colourless, and they are revealed by treating the whole strip with a dye solution – typically Amido Black or Coomassie Blue. Each protein then shows up as a coloured band, so that the whole pattern looks something like a bar code. The thickness of each band and the strength of staining reflect how much of the particular protein is there. What the analyst looks for is unusual bands or bands that are uncharacteristically strong or weak. What is also possible (before applying the general protein stain, which tends to be in acid, which will inactivate most proteins) is to apply a specific solution that will show up an individual biologically active protein. This is very effective in showing up enzyme bands, for example. The enzyme LDH has been used as a diagnostic marker in this way because most of our tissues contain varying proportions of five isoenzymes (see Appendix 10) of LDH, which separate very clearly in electrophoresis (Fig. A15.1). Heart muscle, skeletal muscle and liver will each have distinctively different patterns of the five bands, and this can be helpful in indicating myocardial infarct, for instance.

So-called 'activity staining', as described above, has been used to show up the enzyme bands in Fig. A15.1. It can be seen that, in the native state (unlike Fig. A15.2 where detergent is used to unfold the protein molecules), isoenzymes differ from one another in charge even though they are very similar in size and shape. In the case illustrated in Fig. A15.1, the analysis is not for primary diagnosis but rather to show that cancer cells taken from three breast cancer patients, BC019, 020 and 021, have been successfully cultured without contamination. For comparison L929 is a mouse cell line and HeLa is a very widely used breast cancer cell line taken from an American patient, **Helen Lane**, who died of the disease more than 50 years ago. The point, therefore, is that the three new cancer cultures show very similar isoenzyme patterns to the HeLa cells.

Polyacrylamide gels

The second approach to electrophoresis is to cast a gel (a wobbly jelly held in a tube or now more often between two glass plates) and to use this as the medium through which the current has to flow and through which the proteins therefore have to migrate. The gel acts as a kind of sieve for the large molecules trying to fight their way through: the small molecules can get through easily but large ones may have

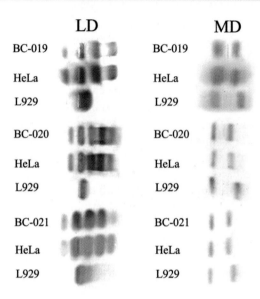

Figure A15.1 Separating isoenzymes by electrophoresis. In the native state (i.e. unlike Fig. A15.2 where detergent is used to unfold the protein molecules), isoenzymes differ from one another in charge even though they are very similar in size and shape. After separation they can be revealed by making use of their catalytic activity and linking the production of NADH to a colour reaction. The image is from Chao Shen *et al.* in *Cancer Cell International* 9:2 (2009) doi:10.1186/1475-2867-9-2. The left panel (LD) shows the separation of the five common isoenzymes of lactate dehydrogenase and the right panel (MD) shows similarly the separation of the two main forms of malate dehydrogenase (a Krebs cycle enzyme).

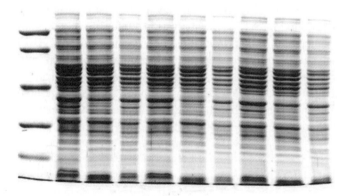

Figure A15.2 SDS-PAGE. One lane contains a set of marker proteins of known M_r so that the M_r value can be estimated for any other protein band on the same gel. Proteins are mostly colourless but can be revealed by staining with a dye such as Coomassie Blue and then destaining to remove excess dye from the background to give a good contrast.(By courtesy of Enzolve Technologies, Dublin).

a struggle. Starch gels used to be popular for this purpose, but in recent times the most common gel to be used is polyacrylamide, and you are likely to come across PAGE gels and SDS-PAGE. PAGE stands for polyacrylamide gel electrophoresis. SDS stands for sodium dodecyl sulphate, which is a detergent. This reflects the fact that there are two ways, both useful, of running PAGE gels. So-called 'native gels' do not involve any detergent. The proteins separate in the way we have already described above, except that now sieving according to the size of the native folded protein may play a larger role, along with the net charge. As before, the proteins can be revealed with a general protein stain, but, since they have been separated in their 'native' (i.e. normal and active) state, they can also be stained for specific biological activity. SDS, on the other hand, being a detergent, unfolds the native structure of proteins, so that they are all in solution like random, microscopic pieces of spaghetti coated with detergent. This might seem like a bad idea, since it throws away the possibility of specific activity staining. However, SDS-PAGE has become an extraordinarily widely used technique because it was realised that the detergent also masks the individual surface charge so that all the different proteins move strictly according to their size – large ones slowly and small ones fast. A gel run can be calibrated with marker proteins of known M_r (Fig. A15.2).

APPENDIX 16

Chromatography and mass spectrometry to separate and identify metabolites

As mentioned in Topic 53, for many metabolites, as for enzymes and other proteins, there are specific methods for detection and measurement. For example, to measure elevated levels of the amino acid phenylalanine in newborn babies with PKU, there are specific enzyme-based kits to measure the concentration of that compound and nothing else. However, there are many metabolites for which similar committed tests are not available, and also such methods are only useful when you know precisely what you are looking for. As with proteins (Appendix 15), there is also a need for a general survey technique that will signal abnormal levels of compounds whatever they are. In this regard, various forms of chromatography are widely used and powerful methods. The word 'chromatography' is a bit misleading, because it implies separating and visualising coloured things. This is because the first application of the method, back in the nineteenth century, was to do exactly that – separating plant pigments. Nowadays, the term is used in a much more general way to describe any separation using the chromatography approach regardless of whether the things being separated happen to be coloured or not. The basic idea is that you have a stationary phase and a mobile phase, and the things you are trying to separate can either stay moving along in the mobile phase or else linger a while attached to the stationary phase. The mobile phase might be liquid (liquid chromatography, LC) or gas (gas chromatography, GC). The stationary phase may be solid or sometimes a coating of liquid on solid particles or on a surface (gas–liquid chromatography, GLC). As a rule, the starting point is that the sample to be separated is in the stationary phase: in LC the sample will be dissolved in the liquid mobile phase, and in

Pain-Free Biochemistry   Paul C. Engel
© 2009 John Wiley & Sons, Ltd

GC the sample will be volatile (i.e. easily vaporised) and it will be injected into a heated stream of the mobile gas phase.

Why should the components in the sample leave the mobile phase to dwell on the stationary phase? There are a wide range of commercially available stationary phases for chromatography and they make use of various properties as the basis for separation, but two of the most important are charge and hydrophobic interaction. Taking charge first, this is the basis of ion-exchange chromatography. The solid support, usually in the form of a powder or perhaps small beads, is charged and typically will be poured as a slurry into a long vertical tube or 'column'. If, say, it is negatively charged, then anything in the mobile phase that has a positive charge will be attracted to the stationary surface. Dissolved positively charged molecules will then tend to separate because those that are very strongly positive will tend to spend most of their time on the stationary phase, while those that are only weakly charged will spend more of their time in the mobile phase. Some substances will be so strongly attracted to the solid phase that they do not move at all. These, however, can be progressively chased off the stationary phase by gradually increasing the concentration of (e.g.) salt (sodium chloride) in the mobile phase, offering the attraction of negative chloride ions in solution to rival the appeal of the solid phase.

If the separation is based on hydrophobic attraction, then the stationary phase must present a surface with a good deal of hydrophobic character, and now the components in the mobile phase will tend to separate according to how hydrophobic they are.

In either case, as the liquid phase comes out of the bottom of the column one collects successive 'fractions' to keep apart the components that have been separated, and these fractions can be separately analysed.

A major development in chromatography was the introduction of high-performance LC (HPLC). This could equally be called high-pressure LC because the separation takes place as the liquid is pumped through the column under high pressure. The result of this is that the separation is very fast, typically taking minutes rather than hours. The advantage of this speed is not simply that it saves time; if a sample spends a long time on the chromatography column, as well as moving steadily down and separating, the components will tend gradually to spread. This to some extent undoes the good work of the chromatography. Hence, HPLC offers the advantage of very clean, sharp separations, because the peaks of separated components have not had time to spread (Fig. A16.1).

One of the standard procedures with chromatographic separations by HPLC or GC is to run known standards down the column under exactly the same conditions, so that one can see whether an unknown component comes off the column after exactly the same time (or in exactly the same volume) as one of the standards. However, this always leaves a small element of doubt – might there be two compounds that just happen to come off the column similarly? In sophisticated analytical systems, this doubt can be resolved: the output from the HPLC or GC can be injected directly into the input port of another instrument, the mass spectrometer. There are various types of mass spectrometry, but one of the most revealing modes smashes

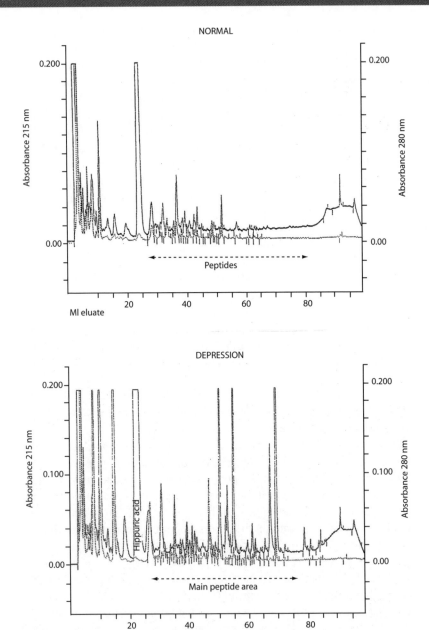

Figure A16.1 HPLC separation of components in a mixture. The chromatograms are from Ying Liu *et al.*, *Behavioral and Brain Functions* 3:47 (2009) doi.1186/1744-9081-3-47. They show analysis of morning urine samples from a normal subject (top) and a severely depressed female patient (below). The analysis is looking for significant differences in the content of peptides in the urine. The samples in such analysis are injected into the HPLC column and as the samples passes down the column the components separate so that they emerge (left to right) as separate sharp peaks, which can be detected by their u.v. light absorption properties. The difference between the normal subject and the patient is striking.

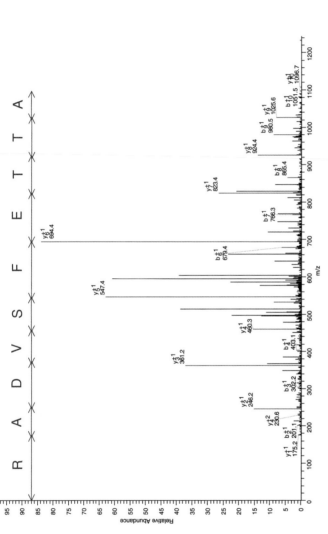

Figure A16.2 Mass spectrometry. There are various types of mass spectrometry but one of the most revealing smashes a molecule into fragments which are separated according to their sizes, which are precisely identified. The pattern of fragments usually identifies the original compound unambiguously. They are separated according to their charge-to-mass ratio. Although to start with, the mass of each fragment is unknown, we do know that the charge has to be a whole number, 1, 2, 3, etc. and the whole story must fit together and make sense. In this particular example, an eleven-residue peptide from a human protein is separated from everything else and then fragmented along the peptide backbone (single splits), which can occur at each peptide bond. This gives two series of fragments with the positive charge either on the C-terminal side of the split (blue) or the N-terminal side (red). Distances between adjacent lines in each series correspond to the mass of an amino acid residues and allow us to read the sequence (single letter codes above, A for alanine, S for serine, R for arginine, etc.). In theory this can all be done by hand but in practice computer software comes up with the unique answer to the puzzle. (Data by courtesy of Dr Giuliano Elia, University College Dublin).

up molecules into a variety of smaller fragments, and the fragmentation pattern usually identifies beyond reasonable doubt what the molecule is. There might be half a dozen different possible compounds sharing the same Mr value, but once you see how the molecule breaks up, this is likely to be consistent with only one of the possibilities (Fig. A16.2).

The instrumentation is expensive and requires a skilled operator, but the power of the method has led to its introduction, for example in a number of paediatric departments for neonatal screening to pick up and identify abnormal metabolites.

Glossary

Acetyl group CH_3CO–

Acid Any substance readily able to shed a proton, H^+.

Active site The region of an enzyme molecule directly involved in handling the reactants.

Aerobic Describes metabolism that uses available oxygen.

Aldehyde A chemical compound carrying a –CHO group.

Allosteric regulation A mechanism for increasing or decreasing an enzyme molecule's activity by substances attaching to positions different from the active site.

Amino acid A substance with a carboxyl group and an amino group. Twenty different amino acids form the building blocks of proteins.

Amino group $-NH_2$

Amphipathic A description of a substance that has both hydrophobic and hydrophilic properties in different parts of its molecule.

Amyloid An insoluble aggregation of misfolded protein, e.g. plaques in brain cells of Alzheimer's patients.

Anabolism Metabolism that converts small building blocks into bigger molecules.

Anaerobic Describes metabolism that does not use oxygen.

Antibiotic Any compound that attacks infecting organisms (usually bacteria). May be either a natural product or synthetic.

Antibody A molecule produced by the immune system and able specifically to recognise and attach to a 'foreign' molecule or object.

Pain-Free Biochemistry Paul C. Engel

Anticodon   The set of three RNA bases on the tip of each tRNA molecule that recognises the complementary codon on an mRNA strand, allowing translation into amino acid sequence.

Apoptosis   The process of programmed cell death that is part of the remodelling of a growing or developing tissue.

Atom   The smallest unit of a chemical element.

Atomic number   A number equal to the number of protons in the atom, which also identifies the chemical element.

ATP   Adenosine 5′-triphosphate (the universal energy currency of the cell).

Avogadro's number   The number of molecules in a mole, e.g. in 12 g of carbon.

Bacterium (plural bacteria)   A small, single-cell organism without a nucleus. Various bacteria cause infections, but others are either harmless or beneficial.

Base   Two meanings

1 Chemical term for a substance that can pick up a proton, H^+, or shed a hydroxide ion, OH^-.

2 A collective term for adenine, guanine, cytosine, thymine and uracil, which are part of the structure of nucleic acids.

Beta-oxidation   The major metabolic pathway for oxidising fatty acids, so-called because in each cycle it is at the beta-carbon (two along from the carboxyl group linked to CoA) where oxidation occurs.

Biosynthesis   = anabolism – see above.

Calorific   Energy yielding.

Carbohydrate   A sugar or a substance made up of linked sugar units.

Carbonyl group $\rangle C{=}O$

Carboxyl group $-COOH$. Normally ionised at physiological pH and conferring acidic properties.

Carcinogen   Any chemical substance that induces cancer.

Catabolism   Metabolic breakdown of big, complex molecules into smaller fragments.

Catalyst   A chemical substance that, in small amounts, can greatly speed up a reaction involving other substances.

Cellulose   A linear polysaccharide made up of glucose units but indigestible for humans. Makes up plants' cell walls.

Cholesterol A lipid substance that we both make and take in via our diet. Important component of membranes, lipoproteins, etc. but in excessive levels contributes to atherosclerotic plaques. Precursor of steroid hormones.

Chromosome One of the 23 pairs (in human beings) of structures in the cell nucleus, each carrying hundreds of genes and packed with protein.

Chylomicron A large lipoprotein particle made in the intestinal lining cells to package digested fat for transport round the body.

CoA Coenzyme A, a B-vitamin-derived cofactor essential for fatty acid metabolism, pyruvate oxidation, TCA cycle, etc.

Codon A sequence of three nucleic acid bases that tells the protein synthetic machinery to insert a particular amino acid in a growing protein (or else to start or stop).

Coenzyme Usually an organic cofactor (see below) that participates in an enzyme-catalysed reaction.

Cofactor A small helper molecule needed to assist an enzyme protein do its job.

Compartmentation The division of a cell into separate specialised locations for different processes and pathways.

Concentration The amount of a substance in a given volume.

Covalent The term describing a tight chemical link between atoms based on sharing electrons.

Cytochrome Several cytochromes, b, c, a, etc., are haem-containing membrane proteins that form part of the mitochondrial respiratory chain.

Cytosol The fluid content of a cell surrounding its organelles.

Dominant mutation A change in a gene such that it only needs to happen in one of the two copies (maternal and paternal) in order to produce a characteristic change, e.g. an inherited disease, hair colour, etc.

Electron One of the three sub-atomic particles that make up all atoms. Negative charge, very little mass (weight).

Element One of the $\sim$100 different fundamental chemicals (each with a symbol C, H, N, etc.) whose atoms in various combinations make up all substances.

ELISA Enzyme-linked immunosorbent assay. A cunning way of combining the recognition specificity of an antibody with the catalytic power of a chemically attached enzyme to give very sensitive measurement.

Enzyme A biological catalyst (see above). Usually very potent and very selective.

Equation   A mathematical term, but also used to describe the statement of the combining ratios for the reactants and products in a chemical reaction.

Equilibrium   The situation where a chemical reaction has come to rest, going no further forward or backward.

Fatty acid   A substance with a long (16–20 carbons) hydrocarbon chain and a carboxyl group at one end. A major building block of fat (lipid) molecules.

Formula   The statement in atomic symbols of the composition of a molecule, e.g. H_2O for water.

Futile cycle   A sequence of reactions that, if uncontrolled, would go round and round, wasting the stored chemical energy of ATP.

Gene   The molecular instructions, encoded in nucleic acid sequence, for making a protein. In human cells the genes are on long DNA molecules in the nuclei.

Genetic code   The key that governs what nucleic acid codon specifies what amino acid goes into a growing protein chain.

Genome   The entire genetic content of a cell.

Gluconeogenesis   The process of making glucose from other metabolites such as lactate, glycerol and amino acids.

Glutathione   A tripeptide with an —SH group. By being oxidised, linking in pairs via a disulphide bond, glutathione molecules form an important part of our defences against reactive oxygen species.

Glycogen   A storage polysaccharide in muscle and liver made of hundreds of glucose units joined to form tree-like molecules.

Glycolysis   The metabolic pathway leading from glucose to pyruvate or lactate.

Glycosidic bond   The chemical (ether) link between two sugar units (e.g. in glycogen or sucrose).

Half-life   The time taken for a chemical decay process, e.g. radioactive decay, to go halfway.

Hexose   Any six-carbon sugar.

Hormone   A chemical messenger between one part of the body and another.

Hydrophilic   Description of a chemical substance or group that readily associates with water molecules.

Hydrophobic   Description of a chemical substance or group that avoids water molecules.

Hydroxyl group   —OH, the characteristic group in glycerol, sugars and other biological alcohols.

Hyperglycaemia   A state in which the level of blood glucose is above the normal range. The opposite is hypoglycaemia, in which the level of blood glucose is below the normal range.

IgG   A class of antibody molecule in the bloodstream.

Immunity   The ability of living things to recognise foreign substances and organisms and target them for destruction.

Inhibition   Slowing down or opposing a process, e.g. an enzyme-catalysed reaction.

Ion   An atom that has either gained or lost electrons, thus attaining a stable outer electron shell but also acquiring a net electrical charge either negative or positive.

Isoenzyme   One of two or more different versions of the same enzyme, i.e. different protein molecules catalysing the same reaction.

Isotope   One of two or more forms of the same chemical element differing only in the number of neutrons in the nucleus and therefore in relative atomic mass, e.g. carbon 12, 13, 14.

Ketogenesis   The production of ketone bodies.

Ketone   A carbon compound with a carbonyl group in the middle, rather than at the end.

Ketone body   Acetoacetate or β-hydroxybutyrate, soluble and easily transported metabolites produced from fatty acids and some amino acids.

Kinase   A type of enzyme that catalyses transfer of phosphate from ATP or GTP to another substance.

Krebs cycle   A circular sequence of eight reactions, starting and ending with oxaloacetate, which oxidises CH_3CO groups (from acetyl CoA) derived from all the major foodstuffs.

Lipase   An enzyme that splits fatty acids off a glyceride.

Lipid bilayer   The basic structure of biological membranes, made up of two back-to-back sheets of phospholipid molecules.

Lipolysis   Mobilisation of fatty acids and glycerol by hydrolysis of triglyceride.

Lysosome   One of the types of intracellular organelles, containing digestive enzymes in an acidic environment, like a cellular stomach.

Membrane   A sheet structure surrounding individual cells and also their internal organelles.

Metabolism   The entire network of biochemical reaction pathways that sustains life.

Methyl group $-CH_3$

Micelle   A ball-like structure formed when amphiphilic molecules, e.g. detergents or phospholipids, bury their hydrophobic structures together away from the surrounding aqueous (watery) environment.

Mitochondrion (plural mitochondria)   Small organelle, plentiful in aerobic cells, and the site of pyruvate oxidation, TCA cycle, respiratory chain and fatty acid β-oxidation.

Molar   An expression of the concentration of a substance in terms of the number of moles of it in a litre.

Mole   A weight of a substance equal to 1 g times the relative molecular mass (e.g. for water $M_r = 18$ and 1 mole = 18 g).

Molecule   The smallest unit of a chemical substance.

Mutation   A change in a cell's DNA that is passed on to succeeding generations.

Neutron   One of the three sub-atomic particles; no charge but mass approximately equal to that of a proton.

Nucleus   Two quite separate meanings:

1 The positively charged centre of an atom containing protons and neutrons.

2 The large organelle in a eukaryotic (e.g. human) cell that contains the chromosomes.

Oxidation   Chemical reaction that adds oxygen or removes hydrogen or removes electrons to increase valency.

Oxidative phosphorylation   Process of ATP formation from ADP and phosphate in mitochondrion, driven by respiratory chain oxidation of NADH and $FADH_2$.

Pentose   Any five-carbon sugar.

Peptide bond   The linkage between one amino acid unit and the next in a protein molecule, formed between a carboxyl and an amino group.

pH   A number that indicates how acid or alkaline a solution is. Defined as minus log $[H^+]$ so that, e.g. if $[H^+] = 10^{-5}$ M then the pH is 5.

Pharmacogenomics   A modern way of approaching drug therapy to take account of relevant genetic differences from patient to patient.

Phosphatase   An enzyme that catalyses removal of a phosphoryl group from, for example, a protein or a sugar.

Phospholipid   A category of lipid (fat) with amphiphilic molecules, usually because two of the three positions on a glycerol molecule carry fatty acids (hydrophobic) and the third position has a hydrophilic substituent.

Phosphorylation   Transfer of a phosphoryl group ($-PO_3^{2-}$), e.g. from PEP to ADP to make ATP or from ATP to glucose to make G6P.

Polymorphism   One of several normal variants across the human population at a particular position in a gene or protein.

Polypeptide   A long chain of amino acids joined to one another via peptide bonds.

Polysaccharide   A carbohydrate with a molecule comprising large numbers of linked sugar units.

Primary structure   Referring to a protein, the precise sequence of the amino acids making up the polypeptide chain.

Prion   kind of protein molecule believed to be the infectious agent for BSE, scrapie in animals, etc.

Proteasome   An intracellular structure that breaks down protein molecules that have been tagged for destruction and recycling of the amino acids.

Protein   A category of diverse biological macromolecules made up of linked amino acid units.

Proteinase   Any one of several different types of enzymes that breaks down proteins by splitting peptide bonds.

Proteolysis   The process of breaking proteins down into smaller peptides and ultimately amino acids.

Proton   One of the three sub-atomic particles; positive charge and mass approximately equal to that of a neutron.

Quaternary structure   The arrangement of subunits in a multi-subunit protein.

Radioimmunoassay   A sensitive technique for measuring biological levels of a substance by using a specific antibody and the substance in radioactive form. Originally introduced to measure hormone levels.

Recessive mutation   A mutation that only makes its presence felt if both inherited copies are similarly affected.

Receptor   A protein molecule, usually sitting in a membrane, which recognises a specific chemical, e.g. a hormone or a neurotransmitter, and passes on a physiological message.

Reduction   Chemical reaction that removes oxygen or donates hydrogen or donates electrons to decrease valency.

Relative molecular mass (Mr)   A number indicating how many times heavier a molecule is than the smallest atom, hydrogen, which is assigned a value of 1.

Replication   Copying. In biochemistry the term usually refers to the process by which a double-stranded DNA molecule gives rise to two identical copies.

Repressor   A substance that keeps a particular gene switched off unless it is removed (by an inducer).

Respiratory burst   A specialised oxidative process in phagocytic blood cells, generating reactive oxygen species to assist in the destruction of engulfed cells, e.g. bacteria.

Respiratory chain   A sequence of substances in the inner membrane of mitochondria that hand on reducing equivalents (from NADH or $FADH_2$) one to another and finally to oxygen.

Restriction enzyme   An enzyme that cuts long double-stranded DNA chains only occasionally, where it finds a specific sequence of nucleotide bases. Physiologically a defence against foreign DNA; scientifically a tool for molecular genetics.

Rhesus factor   One of the genetically determined factors on the surface of blood cells. Can be responsible for immune rejection of a foetus in pregnancy.

Ribosome   The intracellular machine for reading mRNA molecules and producing the new protein molecules. Made up of RNA and proteins. A cell typically has a large number of ribosomes.

Salt   Commonly denotes sodium chloride, but in chemistry is used more generally to describe any similar compound formed by positive metal ions partnering negative ions, e.g. potassium bromide, magnesium chloride, copper sulphate, etc.

Second messenger   A chemical substance, e.g. cAMP, produced inside a cell in response to the arrival of a hormone (the first messenger) at its receptor on the outer surface of the cell membrane.

Secondary structure   Regular repeating sub-structure in a protein molecule.

Specificity   The property of an enzyme, hormone, antibody, receptor, etc. to be extremely selective in recognising and attaching to another substance.

Starch   Plant-derived, digestible mixed carbohydrate made up of linked glucose units.

Statin   A class of drugs that work by inhibiting the enzyme HMG CoA reductase, which controls the synthesis of cholesterol.

Sugar   The basic unit of carbohydrates. Must have at least three carbon atoms, one being a carbonyl group, the others carrying −OH groups.

TCA cycle   See Krebs cycle.

Tertiary structure   The folding pattern giving a protein its 3-D shape.

Transamination   A process in both the breakdown of amino acids and the synthesis of new ones, in which amino groups are transferred between different ketoacids.

Transfer RNA   A specialised kind of RNA molecule about 80 bases long, which recognises a codon in mRNA and feeds the corresponding amino acid to the growing protein molecule via the ribosomal machinery.

Transcription   The process by which the genetic message in DNA is used to produce multiple complementary mRNA molecules to carry the message to the ribosomes.

Translation   The process by which ribosomes use an mRNA message to specify the amino acid sequence of a new protein molecule.

Triglyceride   The most abundant form of fat, either stored or in food. Each molecule has three fatty acids linked to glycerol via its three −OH groups.

Triose   A three-carbon sugar, dihydroxyacetone or glyceraldehydes.

Triplet code   A description of the genetic code emphasising that it takes three DNA/RNA bases to specify one amino acid in a protein sequence.

Uncoupling agent   A substance that breaks the linkage between respiratory oxidation–reduction processes and the formation of ATP.

Unsaturated   Usually refers to fatty acids with double bonds in their hydrocarbon chains.

Urea cycle   The circular metabolic sequence in liver to convert excess amino groups into a safe excretion product.

Valency   A number indicating the combining power of the atoms of an element, e.g. 1 for H, 2 for O, 4 for C, etc.

Virus   A very stripped-down life form that may only have a protein coat and a very few genes inside but is able to take over a cell of a higher organism and switch its machinery over to making new virus particles. Viruses are responsible for many troublesome/dangerous infections.

Vitamin   An organic chemical substance required regularly in trace amounts in the diet to maintain health by allowing us to make essential enzyme cofactors.

X-linkage   Refers to a number of conditions where a defect in a gene on the X-chromosome only affects males because they only have one X-chromosome. In females a good second copy compensates.

Zymogen   An inactive precursor form of an enzyme produced so that activity is only switched on when it is needed, e.g. in digestion or blood clotting.

Answers to self-test MCQs

Chemistry I	1d; 2a; 3c	Topic 26	1b; 2a
Chemistry II	1d; 2c; 3c	Topic 27	1c; 2d
Chemistry III	1c; 2d; 3a; 4c; 5f	Topic 28	1b; 2d
Topic 3	1d; 2a	Chemistry X	1c; 2b; 3c
Topic 4	1b; 2c	Topic 29	1a; 2d; 3b
Topic 5	1d; 2b; 3c	Topic 30	c
Topic 6	1b; 2a; 3b	Topic 31	1a; 2b
Topic 7	1d; 2a; 3d	Topic 32	1c; 2c
Topic 8	b	Topic 33	1a; 2b; 3a
Chemistry IV	1c; 2d, f, g	Topic 34	1d; 2c; 3c; 4a
Topic 9	1d; 2b	Topic 35	1a; 2c; 3d
Chemistry V	1a; 2c	Topic 36	1d; 2c; 3d
Topic 10	1a; 2a	Topic 37	a
Chemistry VI	1c; 2a	Topic 38	e
Topic 11	1d; 2b	Topic 39	b
Topic 12	1c; 2a	Topic 40	1c; 2d; 3a
Topic 13	1b; 2d; 3c	Topic 41	1c; 2a
Topic 14	1d; 2c; 3a	Topic 42	d
Topic 15	1a; 2b; 3d; 4b	Topic 43	1d; 2c
Topic 16	1c; 2b	Topic 44	d
Chemistry VII	1a; 2d; 3c	Topic 45	a
Topic 17	1b; 2c	Topic 46	1b; 2d
Chemistry IX	a	Topic 47	1b; 2c; 3a
Topic 18	1c; 2b	Topic 48	1a, d, f; 2b
Topic 19	1c; 2d; 3a	Topic 49	1c; 2a; 3d; 4a
Topic 20	1c; 2a; 3b; 4d	Topic 50	1d; 2c; 3d; 4a; 5d
Topic 21	1c; 2a; 3d	Topic 51	1b; 2a; 3d; 4b
Topic 22	1c; 2b	Topic 52	1c; 2d; 3d
Topic 23	1a; 2c; 3d	Topic 53	1a; 2a; 3d
Topic 24	1b; 2a; 3a	Topic 54	1b; 2a; 3c; 4a
Topic 25	c		

Pain-Free Biochemistry   Paul C. Engel
© 2009 John Wiley & Sons, Ltd

Index